职业教育赛教一体化课程改革系列规划教材

物联网移动应用开发

WULIANWANG YIDONG YINGYONG KAIFA

周雯 胡荣 主编
郭波涛 黄涛 张志 郭瑞 副主编

中国铁道出版社有限公司
CHINA RAILWAY PUBLISHING HOUSE CO., LTD.

内 容 简 介

物联网移动应用开发是物联网应用技术专业与软件技术专业的重点专业课程,本书采用"Android Studio 应用程序"开发整个系统,旨在使读者掌握物联网应用系统开发中的思路、方法和常用技术。本书吸纳一线教师的教学经验和企业成熟的开发成果,具有通俗易懂、内容精炼、重点突出、层次分明、实例丰富的特点。通过本书的学习,读者可以具备使用 Android 进行物联网应用系统代码编写、修改、测试的能力,可以从事 Android 软件开发工程师、软件测试工程师、系统维护工程师等具有广阔市场前景的工作。

本书从初学者易于理解的角度,以通俗易懂的语言、丰富的实例、简洁的图表将物联网下进行 Android 应用开发的内容介绍给读者。基于唯众物联网实训设备进行讲解,主要内容包括物联网综合实训平台与物联网融合平台的使用、唯众物联网网关操作、温湿度采集、智能楼宇、智慧消防、智能家居、环境数据存储与查询等,最后以唯众物联网融合平台为基础,讲解使用 Android 进行云端开发。

本书适合作为高等职业学校物联网应用技术及软件技术等相关专业的教材。

图书在版编目(CIP)数据

物联网移动应用开发 / 周雯,胡荣主编. —北京:
中国铁道出版社有限公司,2019.8(2022.12重印)
职业教育赛教一体化课程改革系列规划教材
ISBN 978-7-113-25809-2

Ⅰ.①物… Ⅱ.①周… ②胡… Ⅲ.①移动终端—
应用程序—程序设计—职业教育—教材 Ⅳ.①TN929.53

中国版本图书馆 CIP 数据核字(2019)第 155791 号

书　　名:	物联网移动应用开发
作　　者:	周　雯　胡　荣

策　　划:	徐海英	编辑部电话:	(010) 63551006
责任编辑:	王春霞　鲍　闻		
封面制作:	刘　颖		
责任校对:	张玉华		
责任印制:	樊启鹏		

出版发行:中国铁道出版社有限公司(100054,北京市西城区右安门西街8号)
网　　址:http://www.tdpress.com/51eds
印　　刷:北京柏力行彩印有限公司
版　　次:2019年8月第1版　2022年12月第2次印刷
开　　本:787 mm×1 092 mm　1/16　印张:10.5　字数:255 千
书　　号:ISBN 978-7-113-25809-2
定　　价:32.00 元

版权所有　侵权必究

凡购买铁道版图书,如有印制质量问题,请与本社教材图书营销部联系调换。电话:(010) 63550836
打击盗版举报电话:(010) 63549461

前 言

为认真贯彻落实教育部实施新时代中国特色高水平高职学校和专业群建设，扎实、持续地推进职校改革，强化内涵建设和高质量发展，落实双高计划，抓好2019年职业院校信息技术人才培养方案实施及配套建设，在湖北信息技术职业教育集团的大力支持下，武汉唯众智创科技有限公司统一规划并启动了"职业教育赛教一体化课程改革系列规划教材"（《云计算技术与应用》《大数据技术与应用Ⅰ》《网络综合布线》《物联网.NET开发》《物联网嵌入式开发》《物联网移动应用开发》），本书是"教育教学一线专家、教育企业一线工程师"等专业团队的匠心之作，是全体编委精益求精，在日复一日年复一年的工作中，不断探索和超越的教学结晶。本书教学设计遵循教学规律，涉及内容是真实项目的拆分与提炼。

全书内容以实现物联网移动应用开发系统为中心，并适当进行扩展，介绍当前物联网移动应用开发必备的基本技能。项目任务坚持以技能操作培养为中心，遵循理论知识够用的原则组织编写。

本书通过大量的身边易实现的应用案例，帮助学生理解物联网中移动应用开发技术的应用，通过知识拓展开阔视野，培养学生的学习兴趣。

本书作为高等职业学校物联网相关专业学生教材，针对物联网移动应用开发相关领域，结合赛教一体化编写思想，以身边常见生活场景设计的项目为导向来设计任务，使学生在学习之后具备初步的实际工程应用技能。本书也可以作为中职院校、培训机构的物联网专业培训教材，另外对从事物联网、计算机网络的工程技术人员也有一定的参考价值。

本书由武汉软件工程职业学院的周雯、荆州职业技术学院的胡荣担任主编；仙桃职业学院的郭波涛、武汉城市职业学院的黄涛、湖北科技职业学院的张志、武汉唯众智创科技有限公司的郭瑞担任副主编。具体分工如下：武汉唯众智创科技有限公司的郭瑞编写了项目一；武汉软件工程职业学院的周雯编写了项目二；湖北科技职业学院的张志和仙桃职业学院的郭波涛共同编写了项目五；武汉城市职业学院的黄涛编写了项目三；荆州职业技术学院的胡荣编写了项目四。全书由周雯统稿。

由于编者水平所限，加之时间仓促，书中难免出现疏漏或不妥之处，敬请广大读者批评指正。

编 者

2019年5月

目 录

项目一 物联网综合实训平台与物联网融合平台的使用 1
  任务1　实训平台使用 2
  任务2　物联网融合平台使用 6
    任务拓展　通过浏览器获取硬件数据1 11
  任务3　温度采集 13
    任务拓展　通过浏览器获取硬件数据2 18
  任务4　湿度采集 21
    任务拓展　显示平均湿度 25

项目二 智能楼宇 29
  任务1　智能照明 30
    任务拓展　通风控制 36
  任务2　智能考勤 38
    任务拓展　打卡语音播报 46
  任务3　楼宇文化宣传 48
    任务拓展　楼宇室内舒适度显示 53

项目三 智慧消防 59
  任务1　可燃气体数据采集 60
    任务拓展　可燃气语音预警 67
  任务2　智慧消防联动控制 70
    任务拓展　设置阈值1 76
  任务3　智慧消防远程联动 82
    任务拓展　设置阈值2 94

项目四　智能家居	100
任务1　智能门锁	101
任务拓展　RFID与门锁联动	105
任务2　室内光线采集	110
任务拓展　光照与灯光联动	115

项目五　环境数据存储与查询	121
任务1　温度值数据存储	121
任务拓展　SQLite数据库删除和更新	127
任务2　温度值历史查询	135
任务拓展　使用ListView显示数据	147

项目一

物联网综合实训平台与物联网融合平台的使用

项目概述

物联网综合实训平台主要通过物联网关,将移动端与节点终端、物联网执行器等进行物联网连接,主要完成基于物联网、远程访问、设备联动等功能的物联网项目实训。

通过物联网融合平台可以实现对物联网硬件设备的统一管理,实现多网关管理,并生成统一的 API 接口,通过 HTTP 通信方式,实现移动设备对物联网硬件设备数据的收发。

工作任务

- 任务1:实训平台使用。
- 任务2:物联网融合平台使用。
- 任务3:温度采集。
- 任务4:湿度采集。

学习目标

- 理解网关、节点设备、物联网融合平台、Android 设备的通信原理。
- 掌握物联网设备在物联网融合平台中操作的方法。
- 能够通过物联网融合平台生成设备 API 文档。
- 能够通过物联网融合平台的 API 文档获取传感器数据并控制执行器。
- 能够采集温湿度传感器的模拟量数据。

任务 1　实训平台使用

任务描述

学习实训平台中节点终端设备如何获取传感器的数据，并通过节点板上显示屏查看传感器数据。

任务分析

实训平台中，节点终端设备是由无线通信模块、节点底板和传感器构成的，其中无线通信模块主要由 51 单片机、STM32 单片机和 SI4432 芯片组成，它将获取传感器的数据，发送到物联网网关，物联网融合平台与网关通信以获取传感器数据。

知识引入

1. 实训平台设备简介

物联网综合实训平台（见图 1-1）由节点底板（见图 1-2）、无线通信模块（见图 1-3）、传感器（见图 1-4）、继电器（见图 1-5）、网关（见图 1-6）、路由器（见图 1-7）等设备组成。其中，节点底板、无线通信模块可与各类传感器组合成物联节点终端设备。图 1-8 所示为温湿度节点终端设备。

图 1-1　物联网综合实训平台

项目一　物联网综合实训平台与物联网融合平台的使用

图 1-2　节点底板

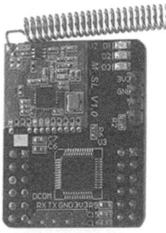

图 1-3　无线通信模块

图 1-4　传感器

图 1-5　继电器

图 1-6　网关

图 1-7　路由器

图 1-8　温湿度节点终端设备

2. 实训平台通信原理

无线通信模块由 51 单片机、STM32 单片机、SI4432 芯片组成。其中，51 单片机负责获取传感器值，然后以串口形式把数值发送给 STM32 单片机，STM32 单片机以 SPI 形式发送给 SI4432 芯片，SI4432 芯片以 443 MHz 的频率将数据以无线方式发给网关，然后移动端 Android 设备就可以通过 Wi-Fi 来和网关通信，以获取传感器的数据值，如图 1-9 所示。

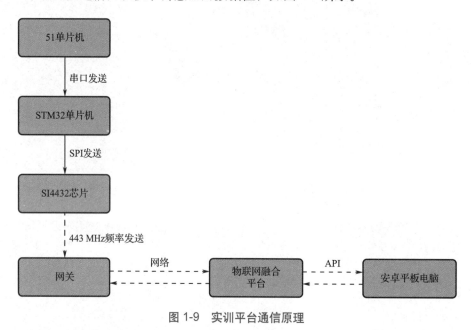

图 1-9　实训平台通信原理

 **任务实现**

（1）设备安装

首先将一个无线通信模块、温湿度传感器安插到工作台的节点底板上，组合成一组物联节点终端设备。然后将网关的网线接入路由器，将电源线接到路由器上，插座插到物联网综合实训平台电源上，再将物联网综合实训平台接到 220 V 的电源上，通电后按下物联网综合实训平台背板上的开关。

（2）单片机程序烧写

将 STC-ISP 下载器（Auto Programmer USB-TTL）延长线接到计算机 USB 接口，引脚线分别插入 RX、TX、GND 及 3 V 引脚，另一头分别接到 51 单片机的 TX、RX、GND 及 3 V 引脚上。需要给模块安装驱动，资料包中提供了两种常用的 USB 串口驱动（CH340 和 CP2102），驱动安装完成后第二个下拉框中选择对应的串口号。串口号可在"计算机管理"→"设备管理器"→"端口"中查看，如图 1-10 所示。

在 STC-ISP 软件（版本号 V6.86C）中第二个下拉列表框里，选择对应的 COM 口，单击"打开工程"按钮，选择"实验 4Slink 温湿度采集实验"中目录 output 下的 .hex 工程文件，检查硬件选项卡里的设置是否与图 1-11 所示一致，单击 STC-ISP 软件界面上的下载程序，按下节点底板的电源键，等待进度条运行完成，此时程序烧写完成。

项目一　物联网综合实训平台与物联网融合平台的使用

图 1-10　查看串口号

图 1-11　STC-ISP 软件

注意：使用普通的 USB 转 TTL 模块给 STC 单片机下载程序，需要给模块断电，然后再上电，用来触发下载。其中，"断电，再上电"的过程称为热启动。

单片机程序下载成功，进行热启动后，即可在节点板的 LED 显示屏上看到温湿度传感器检测到实时的温度值和湿度值，如图 1-12 所示。

图 1-12 节点板的 LED 显示屏

任务小结

请同学们根据完成情况对完成本次任务的知识、技能等要点进行小结。

任务 1 小结	
知识点掌握情况	
技能点掌握情况	

任务 2　物联网融合平台使用

任务描述

了解物联网融合平台的特点，在物联网融合平台中添加物联网综合实训平台中的节点设备，并生成相关项目的 API 接口。

###

物联网融合平台支持 433 MHz、Zigbee、Lora 等多种协议，支持摄像头、门禁、可视对讲、报警等第三方设备接入，所有接入平台设备生成统一格式的 API，并形成相关文档，为用户在开发中提供帮助。

知识引入

物联网融合平台的特点如下：

① 跨平台：基于 Web 架构，所需的仅仅是网页浏览器或者移动终端，无须局限于使用哪款操作系统，任何可以上网的 PC、智能手机、平板电脑等设备都可以随时随地地访问平台。

② 安全、稳定：系统提供了完善的权限保障机制，平台数据传输身份认证方面采用 MD5 签名验证；对于耗时较为严重，需占用较多资源的功能，实现异步调用、事件驱动模型和事件注册机制来最大限度地发挥异步多线程服务的优点。

③ 技术先进、功能强大：B/S 平台采用 MVC 模式开发。抽象出对象层、展现层和控制层，各层之间没有依赖性，松耦合的代码组织方便进行大规模地并行开发，分批分次对整个系统进行升级、维护、改造提供基础，扩展能力极强。

④ 支持多传感器的规则与动作：平台支持传感器规则定义，根据用户定义的一个或多个条件，后台实时监控，在其满足的情况下，对相应传感器进行控制。

⑤ 多设备管理：平台实现不同类型不同数量的设备管理,如网关、摄像头、门禁、对讲、报警等。

⑥ 提供相关 API：平台根据项目生成对应 API 文档，可供读者在开发时查阅与使用。

任务实现

（1）平台登录

在浏览器中输入物联网融合平台的 IP 地址，在登录界面输入用户名、密码，如图 1-13 所示，然后单击【登录】按钮，进入平台。

图 1-13 平台登录

（2）平台首页会显示当前登录用户的所有项目，如图 1-14 所示。

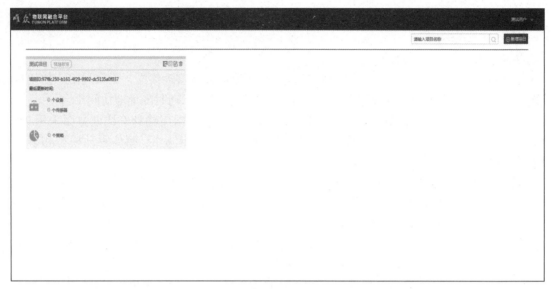

图 1-14　平台首页

如果该用户是第一次登录或将自己项目全部删除，需要创建项目后才可进行后续操作，如图 1-15 所示。

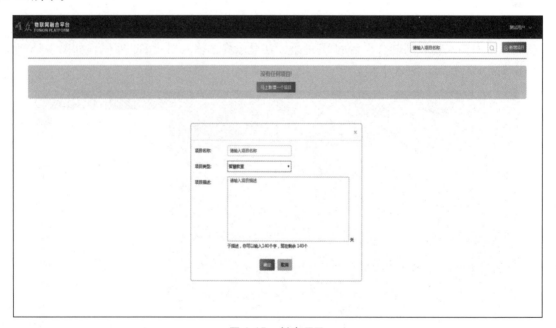

图 1-15　创建项目

（3）创建项目

单击图 1-15 中的【新增项目】或【马上新增一个项目】按钮，在弹出的新增页面中，输入相关信息后单击【确定】按钮，完成项目新增。新增成功的项目会出现在项目列表中，如图 1-16 所示。

项目一　物联网综合实训平台与物联网融合平台的使用

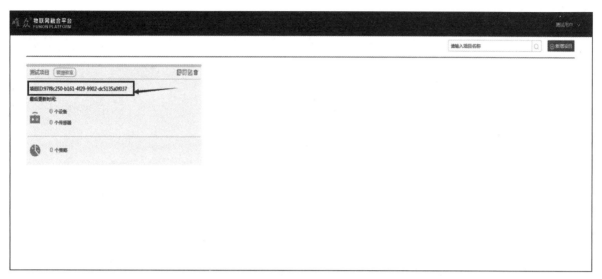

图 1-16　新增项目

一个刚刚新增的项目是没有任何数据的，那么接下来我们需要对项目添加设备，单击上面列表中的设备或传感器数量，跳转到设备管理页面，如图 1-17 所示。

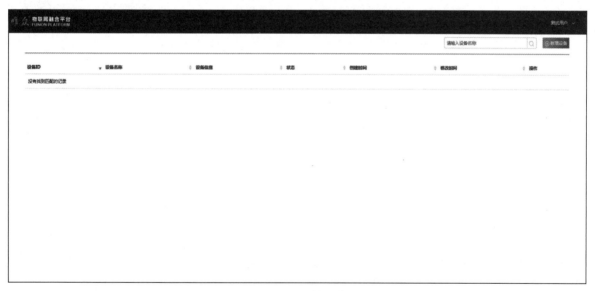

图 1-17　添加设备

在设备管理界面，单击【新增设备】按钮，弹出新增设备界面，在弹出界面中填写设备信息后单击【确定】按钮，如图 1-18 所示，完成新增操作。若新增成功设备会出现在设备列表中，如图 1-19 所示。

图 1-18 新增设备页面

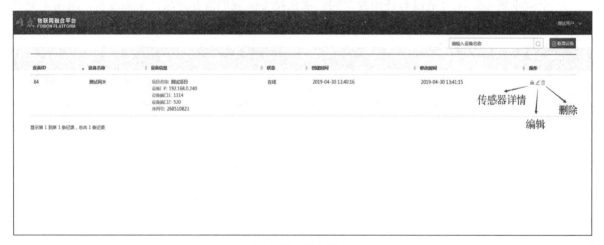

图 1-19 设备列表

注意：
① 添加设备时会验证平台与设备是否连接，如果没有连通或输入信息错误，将新增失败。常见失败原因，以及解决方案如下。
a. 网关未开启（为网关通电，开启网关）。
b. 网关与平台不在同一局域网（设置网关与平台在同一局域网内）。
c. 网关 IP 或 SN 填写错误（检查设备上标签与填写内容）。
② 设备添加成功后，平台会自动解析设备下的传感器，不需要再次对传感器进行添加。
③ 平台定时检查设备是否在线，更新设备列表状态栏中信息，如遇无法获取平台数据的情况，请先检查设备状态。

单击列表操作栏中的按钮对设备进行操作，如图 1-20 所示。

项目一　物联网综合实训平台与物联网融合平台的使用

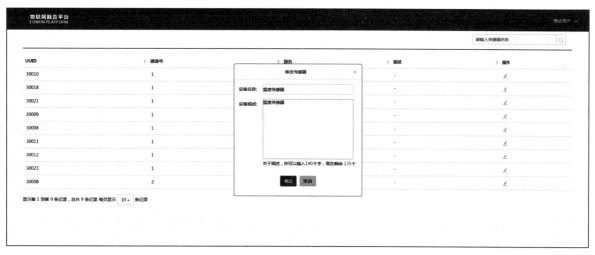

图 1-20　修改传感器

任务小结

请同学们根据完成情况对完成本次任务的知识、技能等要点进行小结。

任务 2 小结	
知识点掌握情况	
技能点掌握情况	

任务拓展　通过浏览器获取硬件数据 1

 任务描述

在浏览器中输入物联网融合平台相关 API 接口，可以获取指定传感器数据。

任务分析

首先在物联网融合平台中生成 API，在浏览器中输入 API 接口，并替换相应内容，如：API 接口如下。

http://192.168.0.195:8080/wziot/wzIotApi/getSensorData/{projectId}/{sn}

用户需将 IP 地址、projectId、sn 进行替换。

任务实现

设备添加完成后，在项目列表界面的项目中单击【发布】按钮，生成该项目的 API 接口，如图 1-21 所示，接下来我们将根据生成的 API 接口进行开发。

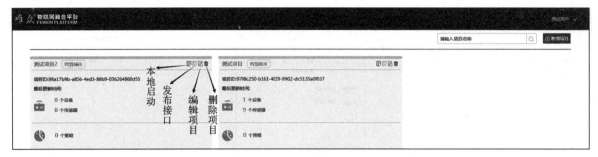

图 1-21　发布项目

单击【发布】按钮后跳转到 API 接口页面，如图 1-22 所示。

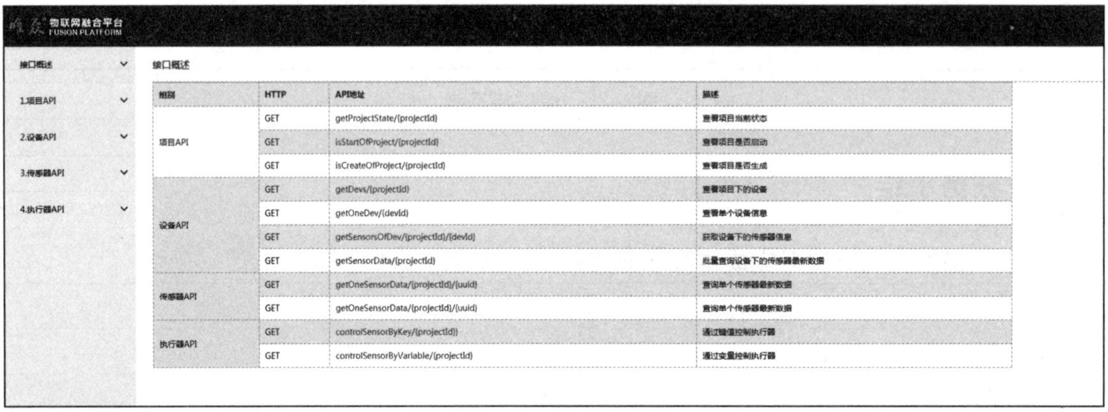

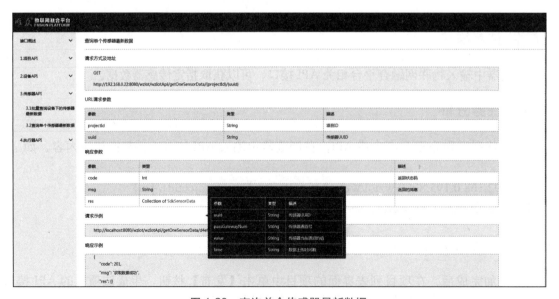

图 1-22　API 接口页面

选择相应接口，单击【查询单个传感器最新数据】，本节以获取温湿度传感器数据为例进行介绍，如图 1-23 所示。

图 1-23　查询单个传感器最新数据

根据 API 生成的接口，将 HTTP 地址复制到浏览器的 URL 地址栏中，并将相关参数进行替换，便可获取到设备的数据或对设备进行控制，如图 1-24 所示。

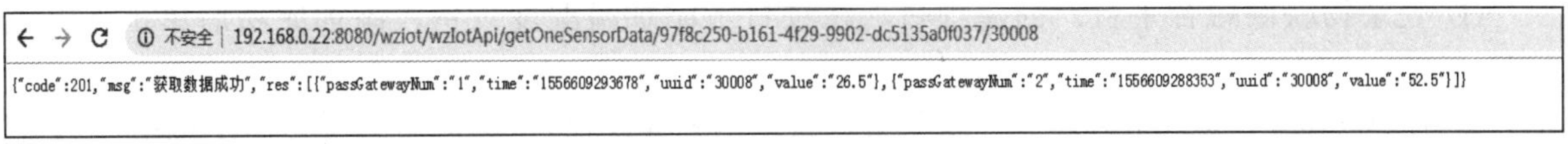
不安全 | 192.168.0.22:8080/wziot/wzIotApi/getOneSensorData/97f8c250-b161-4f29-9902-dc5135a0f037/30008
{"code":201,"msg":"获取数据成功","res":[{"passGatewayNum":"1","time":"1556609293678","uuid":"30008","value":"26.5"},{"passGatewayNum":"2","time":"1556609288353","uuid":"30008","value":"52.5"}]}

图 1-24　HTTP 地址

任务 3　温度采集

任务描述

通过物联网融合平台 API 接口，采集温度传感器的数值，并显示在主界面。

任务分析

首先在物联网融合平台中生成 API，根据对应的 API 获取温度数据，将值显示在页面上。

知识引入

平台 API 使用查询单个传感器最新数据接口（界面见图 1-25），将“请求方式及地址”中 {projectId} 替换为项目的 ID，将 {uuid} 替换为传感器的 ID，可在本书配套资源中查询常用传感器的 ID 号。

例如：项目 ID 为 d4e9f9af-1482-4be9-a096-63f7ae015a03。

温湿度传感器 ID 为 30008。

替换后的 URL 为：

http://192.168.0.193:8080/wziot/wzIotApi/getOneSensorData/d4e9f9af-1482-4be9-a096-63f7ae015a03/30008。

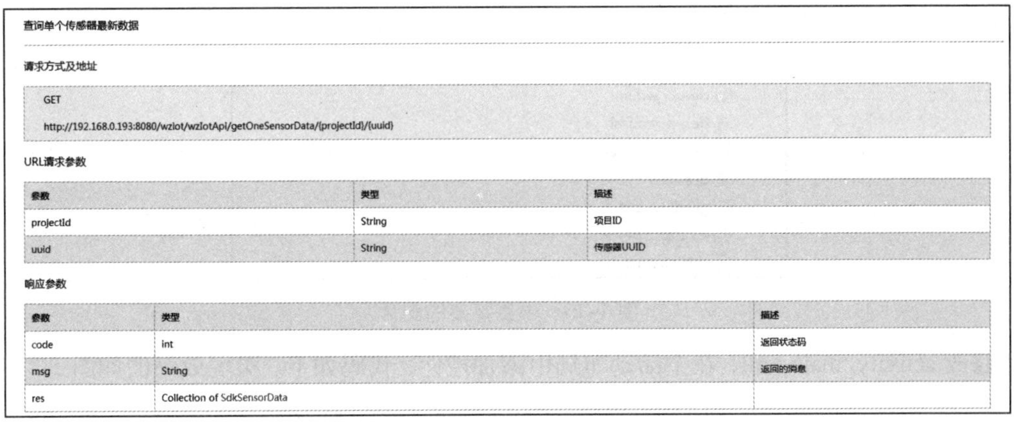

查询单个传感器最新数据

请求方式及地址

GET
http://192.168.0.193:8080/wziot/wzIotApi/getOneSensorData/{projectId}/{uuid}

URL请求参数

参数	类型	描述
projectId	String	项目ID
uuid	String	传感器UUID

响应参数

参数	类型	描述
code	int	返回状态码
msg	String	返回的消息
res	Collection of SdkSensorData	

图 1-25　查询单个传感器最新数据

① 登录物联网融合平台，创建项目，在项目下添加温湿度设备，添加成功后单击【生成API】按钮。

② 在 Android Studio 中创建一个新项目，将应用名称设置为 demo_wd，并为活动添加一个空活动。

③ 将项目需要的图片复制到 res/drawable 文件夹下，如图 1-26 所示。

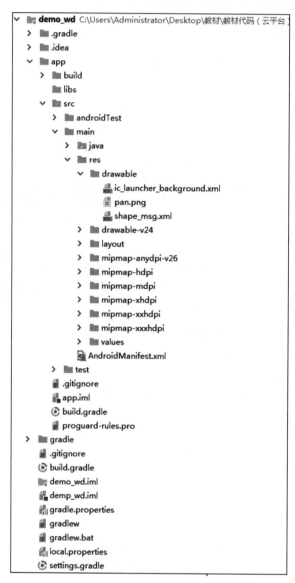

图 1-26　项目需要的图片

④ 修改 activity_main.xml，在主活动布局中添加控件，代码如下，实现效果图如图 1-27 所示。

```xml
<LinearLayout
    android:id="@+id/linearLayout"
    android:layout_width="0dp"
    android:layout_height="0dp"
    android:background="#01123D"
    android:gravity="center"
    android:orientation="vertical"
    app:layout_constraintBottom_toBottomOf="parent"
    app:layout_constraintEnd_toEndOf="parent"
    app:layout_constraintStart_toStartOf="parent"
    app:layout_constraintTop_toTopOf="parent">

    <LinearLayout
        android:layout_width="match_parent"
        android:layout_height="0dp"
        android:layout_weight="2"
        android:gravity="center"
        android:orientation="vertical">

        <LinearLayout
            android:layout_width="wrap_content"
            android:layout_height="wrap_content"
            android:background="@drawable/shape_msg"
            android:gravity="center"
            android:orientation="horizontal"
            android:paddingLeft="25dp"
            android:paddingTop="5dp"
            android:paddingRight="25dp"
            android:paddingBottom="5dp">

            <TextView
                android:layout_width="wrap_content"
                android:layout_height="wrap_content"
                android:text="当前温度"
                android:textColor="@android:color/white"
                android:textSize="30sp" />

        </LinearLayout>
    </LinearLayout>

    <LinearLayout
        android:layout_width="match_parent"
        android:layout_height="0dp"
        android:layout_weight="3"
        android:gravity="center"
        android:orientation="vertical">

        <ImageView
            android:layout_width="450dp"
            android:layout_height="wrap_content"
            app:srcCompat="@drawable/pan" />

        <TextView
            android:id="@+id/svalue"
```

```
                android:layout_width="wrap_content"
                android:layout_height="wrap_content"
                android:layout_marginTop="-240dp"
                android:text="0° "
                android:textColor="#4678E2"
                android:textSize="100sp" />
        </LinearLayout>

</LinearLayout>
```

图 1-27　界面效果图

⑤ 在 AndroidManifest.xml 配置文件中，添加网络访问权限。

`<uses-permission android:name="android.permission.INTERNET"/>`

⑥ 修改 MainActivity.java 文件，创建 initTools（方法）设置请求 API 的 URL 地址，并启动定时器，在 onCreate() 方法中调用，代码如下。

```
private void initTools() {

    // 初始化唯众HTTP请求工具类
    Wz_HttpTools wht=new Wz_HttpTools(handler);
    // 设置请求URL
    wht.setHttpURL("http://192.168.0.193:8080/wziot/wzIotApi/getOneSensorData/
    d4e9f9af-1482-4be9-a096-63f7ae015a03/30008");
    // 创建并启动定时器
    TimerTask task=wht.getJsonData();
    timer.schedule(task, 2000, 2000);
}
```

⑦ 创建 show(String jsonResult, String index) 方法，在该方法中将接收到的温度值显示在界面的文本框中。

```
public void show(String jsonResult, String index) {
```

```
        try {
// 获取 Json 数据中的 JsonArray 数据
            JSONArray obj=new JSONObject(jsonResult).getJSONArray("res");
            for (int i=0; i<obj.length(); i++) {
                JSONObject json=(JSONObject) obj.get(i);
                // 获取通道号
                String rindex=json.getString("passGatewayNum");
                // 判断通道号
                if (index.equals(rindex)) {
                // 获取温度值
                    String value=json.getString("value");
                    svalue.setText(value);
                }
            }
        } catch (JSONException e) {
            e.printStackTrace();
        }
}
```

⑧ 程序运行后，如图 1-28 所示。

图 1-28　温度值读取效果图

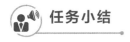任务小结

请同学们根据完成情况对完成本次任务的知识、技能等要点进行小结。

任务 3 小结	
知识点掌握情况	
技能点掌握情况	

任务拓展　通过浏览器获取硬件数据 2

任务描述

通过物联网融合平台 API 接口，采集温度传感器的数值，并显示在主界面，在温度值超过 35℃时，在界面上显示报警标志。

任务分析

首先在物联网融合平台中生成 API，根据对应的 API 获取温度数据，将值显示在页面上；判断温度值，当其大于 35℃时，在页面上显示报警标志，当值恢复到 35℃及以下时，报警标志消失。

任务实现

实现步骤如下。

① 登录物联网融合平台，创建项目，在项目下添加设备，添加成功后单击【生成 API】按钮。

② 在 Android Studio 中创建一个新项目，将应用名称设置为 demo_wd_ex，并为活动添加一个空活动。

③ 将项目需要的图片复制到 res/drawable 文件夹下，如图 1-29 所示。

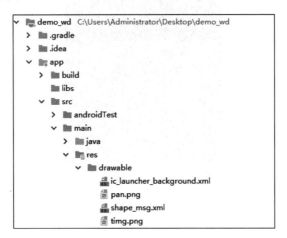

图 1-29　项目需要的图片

④ 修改 activity_main.xml，在主活动布局中添加控件。

```
<LinearLayout
    android:layout_width="match_parent"
    android:layout_height="match_parent"
    android:background="#01123D"
    android:gravity="center"
    android:orientation="vertical">
```

```xml
<LinearLayout
    android:layout_width="match_parent"
    android:layout_height="0dp"
    android:layout_weight="2"
    android:gravity="center"
    android:orientation="vertical">

    <LinearLayout
        android:layout_width="wrap_content"
        android:layout_height="wrap_content"
        android:background="@drawable/shape_msg"
        android:gravity="center"
        android:orientation="horizontal"
        android:paddingLeft="30dp"
        android:paddingTop="5dp"
        android:paddingRight="30dp"
        android:paddingBottom="5dp">

        <TextView
            android:layout_width="wrap_content"
            android:layout_height="wrap_content"
            android:layout_weight="1"
            android:text=" 当前温度 "
            android:textColor="@android:color/white"
            android:textSize="30sp" />

        <ImageView
            android:id="@+id/warn"
            android:layout_width="@android:dimen/notification_large_icon_height"
            android:layout_height="@android:dimen/notification_large_icon_width"
            android:layout_weight="1"
            android:visibility="gone"
            app:srcCompat="@mipmap/ic_launcher"/>
    </LinearLayout>
</LinearLayout>

<LinearLayout
    android:layout_width="match_parent"
    android:layout_height="0dp"
    android:layout_weight="3"
    android:gravity="center"
    android:orientation="vertical">

    <ImageView
        android:id="@+id/imageView"
        android:layout_width="450dp"
        android:layout_height="wrap_content"
        app:srcCompat="@drawable/pan"/>

    <TextView
        android:id="@+id/svalue"
        android:layout_width="wrap_content"
        android:layout_height="wrap_content"
        android:layout_marginTop="-240dp"
```

```
                android:text="0° "
                android:textColor="#4678E2"
                android:textSize="100sp"/>
    </LinearLayout>

</LinearLayout>
```

⑤ 在 AndroidManifest.xml 配置文件中，添加网络访问权限：

```
<uses-permission android:name="android.permission.INTERNET"/>
```

⑥ 修改 MainActivity.java 类，设置请求 API 的 URL 地址，并启动定时器，代码如下。

```
Wz_HttpTools wht=new Wz_HttpTools(handler);
wht.setHttpURL("http://192.168.0.193:8080/wziot/wzIotApi/getOneSensorData/
d4e9f9af-1482-4be9-a096-63f7ae015a03/30008");
TimerTask task=wht.getJsonData();
timer.schedule(task, 2000, 2000);
```

程序运行后获取温度数据，并判断温度值是否超过 35℃，超过则显示报警图片。

⑦ 在 show(String jsonResult, String index) 方法中，将接收到的温度值显示在界面的文本框中，判断温度值是否超标，超标后显示报警图片，代码如下。

```
public void show(String jsonResult, String index) {
    try {
        JSONArray obj=new JSONObject(jsonResult).getJSONArray("res");
        for (int i=0; i < obj.length(); i++) {
            JSONObject json=(JSONObject) obj.get(i);
            String rindex=json.getString("passGatewayNum");
            if (index.equals(rindex)) {
                String value=json.getString("value");
                svalue.setText(value);
                if (Integer.parseInt(value)>35)
                    warn.setVisibility(View.VISIBLE);
                else
                    warn.setVisibility(View.GONE);
            }
        }
    } catch (JSONException | NumberFormatException e) {
        e.printStackTrace();
    }
}
```

⑧ 程序运行后，如图 1-30 所示。

项目一　物联网综合实训平台与物联网融合平台的使用

图 1-30　温度报警效果图

任务 4　湿度采集

任务描述

通过物联网融合平台 API 接口，采集湿度传感器的数值，并显示在主界面。

任务分析

首先在物联网融合平台中生成 API，根据对应的 API 获取湿度数据，获取到的 Json 数据中包含温度与湿度的值，通过通道号找出正确的湿度值，将其显示在页面中。

知识引入

平台 API 使用查询单个传感器最新数据接口（界面见图 1-31），将"请求方式及地址"中 {projectId} 替换为项目的 ID，将 {uuid} 替换为传感器的 ID。

例如：项目 ID 为 d4e9f9af-1482-4be9-a096-63f7ae015a03。

传感器 ID 为 30008。

替换后的 url 为：

http://192.168.0.193:8080/wziot/wzIotApi/getOneSensorData/d4e9f9af-1482-4be9-a096-63f7ae015a03/30008。

查询单个传感器最新数据

请求方式及地址

GET
http://192.168.0.193:8080/wziot/wzIotApi/getOneSensorData/{projectId}/{uuid}

URL请求参数

参数	类型	描述
projectId	String	项目ID
uuid	String	传感器UUID

响应参数

参数	类型	描述
code	int	返回状态码
msg	String	返回的消息
res	Collection of SdkSensorData	

图 1-31 湿度采集数据接口

任务实现

① 登录物联网融合平台，创建项目，在项目下添加设备，添加成功后单击【生成 API】按钮。

② 在 Android Studio 中创建一个新项目，将应用名称设置为 demo_sd，并为活动添加一个空活动。

③ 修改 activity_main.xml，在主活动布局中添加控件，实现效果如图 1-32 所示。

图 1-32 界面效果图

```
<LinearLayout
    android:layout_width="match_parent"
    android:layout_height="match_parent"
    android:background="#01123D"
    android:gravity="center"
    android:orientation="vertical">
```

```xml
<LinearLayout
    android:layout_width="match_parent"
    android:layout_height="0dp"
    android:layout_weight="2"
    android:gravity="center"
    android:orientation="vertical">

    <LinearLayout
        android:layout_width="wrap_content"
        android:layout_height="wrap_content"
        android:background="@drawable/shape_msg"
        android:gravity="center"
        android:orientation="horizontal"
        android:paddingLeft="30dp"
        android:paddingTop="5dp"
        android:paddingRight="30dp"
        android:paddingBottom="5dp">

        <TextView
            android:layout_width="wrap_content"
            android:layout_height="wrap_content"
            android:layout_weight="1"
            android:text=" 当前湿度 "
            android:textColor="@android:color/white"
            android:textSize="30sp"/>

    </LinearLayout>
</LinearLayout>

<LinearLayout
    android:layout_width="match_parent"
    android:layout_height="0dp"
    android:layout_weight="3"
    android:gravity="center"
    android:orientation="vertical">

    <ImageView
        android:id="@+id/imageView"
        android:layout_width="450dp"
        android:layout_height="wrap_content"
        app:srcCompat="@drawable/pan"/>

    <TextView
        android:id="@+id/svalue"
        android:layout_width="wrap_content"
        android:layout_height="wrap_content"
        android:layout_marginTop="-240dp"
        android:text="0"
        android:textColor="#4678E2"
        android:textSize="100sp"/>

</LinearLayout>

</LinearLayout>
```

④ 在 AndroidManifest.xml 配置文件中，添加网络访问权限。

```
<uses-permission android:name="android.permission.INTERNET"/>
```

⑤ 修改 MainActivity.java 类，设置请求 API 的 URL 地址，并启动定时器，代码如下。

```
Wz_HttpTools wht=new Wz_HttpTools(handler);
wht.setHttpURL("http://192.168.0.193:8080/wziot/wzIotApi/getOneSensorData/d4e9f9af-1482-4be9-a096-63f7ae015a03/30008");
TimerTask task=wht.getJsonData();
timer.schedule(task, 2000, 2000);
```

⑥ 在 show(String jsonResult, String index) 方法中，将接收到的数值显示在界面的文本框中，根据传感器 ID 号 "30008" 获取到的 Json 数据中既有温度信息又有湿度信息，需根据通道号判断是温度值还是湿度值，通道号为 1 是温度值，通道号为 2 是湿度值。代码如下。

```
public void show(String jsonResult, String index) {
    try {
        JSONArray obj=new JSONObject(jsonResult).getJSONArray("res");
        for (int i=0; i<obj.length(); i++) {
            JSONObject json=(JSONObject) obj.get(i);
            String rindex=json.getString("passGatewayNum");
            if (index.equals(rindex)) {
                String value=json.getString("value");
                svalue.setText(value);
            }
        }
    } catch (JSONException e) {
        e.printStackTrace();
    }
}
```

⑦ 程序运行后，如图 1-33 所示。

图 1-33　湿度采集运行效果图

项目一　物联网综合实训平台与物联网融合平台的使用

任务小结

请同学们根据完成情况对完成本次任务的知识、技能等要点进行小结。

任务 4 小结	
知识点掌握情况	
技能点掌握情况	

任务拓展　显示平均湿度

任务描述

通过物联网融合平台 API 接口，采集 10 个湿度传感器的值，计算这 10 个值的平均值，并显示在主界面。

任务分析

首先在物联网融合平台中生成 API，根据对应的 API 获取湿度数据，获取到的 Json 数据中包含温度与湿度的值，通过通道号找出正确的湿度值，并采集十个连续的值，计算其平均值。

任务实现

实现步骤如下。
① 登录物联网融合平台，创建项目，在项目下添加设备，添加成功后单击【生成 API】按钮。
② 在 Android Studio 中创建一个新项目，将应用名称设置为 demo_sd_ex，并为活动添加一个空活动。
③ 修改 activity_main.xml，在主活动布局中添加控件，实现的效果图如图 1-34 所示。

图 1-34　界面效果图

```xml
<LinearLayout
    android:layout_width="match_parent"
    android:layout_height="match_parent"
    android:background="#01123D"
    android:gravity="center"
    android:orientation="vertical">

    <LinearLayout
        android:layout_width="match_parent"
        android:layout_height="0dp"
        android:layout_weight="2"
        android:gravity="center"
        android:orientation="vertical">

        <LinearLayout
            android:layout_width="wrap_content"
            android:layout_height="wrap_content"
            android:background="@drawable/shape_msg"
            android:gravity="center"
            android:orientation="horizontal"
            android:paddingLeft="30dp"
            android:paddingTop="5dp"
            android:paddingRight="30dp"
            android:paddingBottom="5dp">

            <TextView
                android:layout_width="wrap_content"
                android:layout_height="wrap_content"
                android:layout_weight="1"
                android:text=" 平均湿度 "
                android:textColor="@android:color/white"
                android:textSize="30sp"/>

        </LinearLayout>
    </LinearLayout>

    <LinearLayout
        android:layout_width="match_parent"
        android:layout_height="0dp"
        android:layout_weight="3"
        android:gravity="center"
        android:orientation="vertical">

        <ImageView
            android:id="@+id/imageView"
            android:layout_width="450dp"
            android:layout_height="wrap_content"
            app:srcCompat="@drawable/pan"/>

        <TextView
            android:id="@+id/svalue"
            android:layout_width="wrap_content"
            android:layout_height="wrap_content"
            android:layout_marginTop="-240dp"
```

```xml
            android:text="0"
            android:textColor="#4678E2"
            android:textSize="100sp"/>
    </LinearLayout>

</LinearLayout>
```

④ 在 AndroidManifest.xml 配置文件中，添加网络访问权限：

```xml
<uses-permission android:name="android.permission.INTERNET"/>
```

⑤ 程序运行后获取湿度数据，并采集 10 个连续的值，计算其平均值，代码如下。

```java
Wz_HttpTools wht=new Wz_HttpTools(handler);
wht.setHttpURL("http://192.168.0.193:8080/wziot/wzIotApi/getOneSensorData/
d4e9f9af-1482-4be9-a096-63f7ae015a03/30008");
TimerTask task=wht.getJsonData();
timer.schedule(task, 2000, 2000);
private LinkedList<Double> data=new LinkedList<>();

public void show(String jsonResult, String index) {
        try {
            JSONArray obj=new JSONObject(jsonResult).getJSONArray("res");
            for(int i=0; i < obj.length(); i++) {
                JSONObject json=(JSONObject) obj.get(i);
                String rindex=json.getString("passGatewayNum");
                if (index.equals(rindex)) {
                    String value=json.getString("value");
                    System.out.println(value);
                    if (data.size() < 10) {
                        data.add(Double.parseDouble(value));
                    } else {
                        data.remove();
                        data.add(Double.parseDouble(value));
                        DecimalFormat df=new DecimalFormat("#.0");
                        svalue.setText(df.format(getSum(data)));
                    }
                }
            }
        } catch (JSONException | NumberFormatException e) {
            e.printStackTrace();
        }

}
private double getSum(LinkedList<Double> data) {
        double sum=0;
        for(Double datum : data) {
            sum+=datum;
        }
        return sum/10;
}
```

⑥ 程序运行后，如图 1-35 所示。

图 1-35 平均湿度运行效果图

项目二 智能楼宇

项目概述

基于物联网构建的智能楼宇，可以使建筑内众多公共资源具有语境感知能力，它使建筑物具有了安全、便利、高效、节能的特点，使其真正成为智慧城市的细胞。基于物联网构建的智能楼宇主要特点如下。

① 智能照明自动控制系统能够节约能源，减少维护费用，改善照明质量。

② 智能考勤采用 RFID 标签对员工进行考勤统计，对进入公司的人员进行身份识别，对合法用户进行考勤统计，对非法用户进行告警。

③ 楼宇文化显示系统借助 LED 电子显示屏进行广告宣传。

工作任务

- 任务1：智能照明。
- 任务2：智能考勤。
- 任务3：楼宇文化宣传。

学习目标

- 理解继电器控制的工作原理。
- 了解 RFID 的技术原理。
- 了解 LED 屏显示的工作原理。
- 能设计开发继电器控制程序。
- 能设计开发 RFID 卡读取的程序。
- 培养学生自主探究和解决问题的能力。
- 培养学生严谨的逻辑思维能力。

任务 1 智能照明

任务描述

楼宇建筑面积一般较大，人工维护烦琐，管理人员希望能将高素质的管理意识运用于照明控制系统中去，将普通照明人为的手动开与关转换成智能化管理，工作人员只需用安卓移动设备就能对灯具进行远程开关，减少公司的人工维护费用。

任务分析

安卓移动设备不能直接控制照明灯，需要一个中间物来协调，也就是单片机，当单片机接收到"开灯"和"关灯"的消息后，再利用继电器（单片机的外接模块），来控制"开灯"和"关灯"。

知识引入

1. 继电器的工作原理

继电器的结构大家可以参考图 2-1。继电器一共有三个输出端口（常开端、公共端、常闭端）。我们可以通过单片机控制继电器的公共端是和常开端连通，还是和常闭端连通。只需要将照明灯的电线的两头分别和继电器的常开端和公共端连接起来即可，如图 2-2 所示。通过单片机控制继电器的公共端和常开端连接时照明灯打开，反之照明灯关闭。

图 2-1 继电器结构　　　　图 2-2 继电器控制台灯原理

整个流程思路：首先使安卓移动设备和单片机的 Wi-Fi 模块建立通信，当安卓移动设备发送一个打开信号时，单片机收到相应的信号并控制继电器的公共端指向常开端，照明灯亮起。

2. Shape 标签

Android 下可以使用 Shape 标签来进行简单 UI 的开发，可以在一定程度上减少图片的使用，降低 App 的体积。一般用 shape 定义的 XML 文件存放在 drawable 目录下，常用属性如下。

① corners: 设置圆角，只适用于 rectangle 类型，可分别设置四个角不同半径的圆角，当设置的圆角半径很大时，比如 200 dp，就可变成弧形了。

— android:radius——圆角半径，会被下面每个特定的圆角属性重写。

② stroke: 设置描边，可描成实线或虚线。

android:color——描边的颜色。

android:width——描边的宽度。

③ solid: 设置形状填充的颜色，只有 android:color 一个属性。

android:color——填充的颜色。

3. 控件的 setEnabled() 方法

setEnabled 使能控件，如果设置为 false，该控件永远不会活动，不论设置为什么属性，都无效；设置为 true，表明激活该控件，控件处于活动状态，能响应事件了，比如触摸、点击、按键事件等。

4. 复选框的 isChecked() 方法

这个方法是用于复选框的，即 CheckBox 对象。区分 CheckBox 是否被选中，isChecked 有两个返回值：当 CheckBox 对象的复选框被选中时，isChecked() 返回 true，即 1；当 CheckBox 对象的复选框没有被选中时，isChecked() 返回 false，即 0。

5. 相关的 API 方法

通过键值控制执行器：

请求方式及地址：http://192.168.0.193:8080/wziot/wzIotApi/controlSensorByKey/{projectId}?uuid={uuid}&key={key}。

URL 请求参数及响应参数如图 2-3 所示。

URL请求参数

参数	类型	描述
projectId	String	项目ID
uuid	String	传感器UUID
key	int	键值

响应参数

参数	类型	描述
code	int	返回状态码
msg	String	返回的消息

图 2-3　URL 请求参数及响应参数

通过变量控制执行器：

请求方式及地址：http://192.168.0.193:8080/wziot/wzIotApi/controlSensorByVariable/{projectId}?uuid={uuid}&index={index}&variable={variable}。

URL 请求参数及响应参数如图 2-4 所示。

URL请求参数

参数	类型	描述
projectId	String	项目ID
uuid	String	传感器uuid
index	String	索引值
variable	int	变量值

响应参数

参数	类型	描述
code	int	返回状态码
msg	String	返回的消息

图 2-4　URL 请求参数及响应参数

任务实现

这里将使用安卓移动设备界面上的按钮和复选框来实现"打开"或"关闭"照明灯的功能，具体操作步骤如下。

① 登录唯众物联网融合平台，创建项目，在项目下添加设备，添加成功后单击【生成API】按钮。

② 在 Android Studio 中创建一个新项目，将应用名称设置为 demo_dg，并为活动添加一个空活动。

③ 将项目相关的图片复制到 res/drawable 文件夹下，如图 2-5 所示。

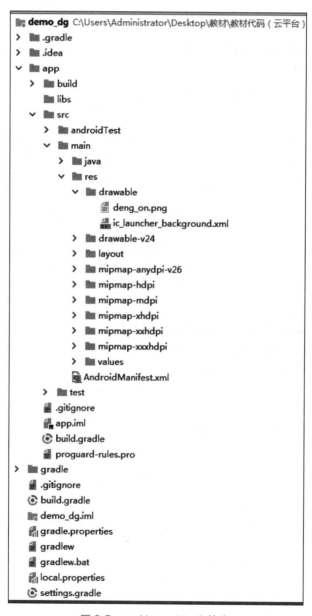

图 2-5　res/draw able 文件夹

④ 修改 activity_main.xml，在主活动布局中添加控件，代码如下，实现效果图如图 2-6 所示。

```xml
<LinearLayout
    android:layout_width="match_parent"
    android:layout_height="match_parent"
    android:background="#01123D"
    android:gravity="center"
    android:orientation="vertical">

    <LinearLayout
        android:layout_width="match_parent"
        android:layout_height="0dp"
        android:layout_weight="2"
        android:gravity="center"
        android:orientation="vertical">

        <LinearLayout
            android:layout_width="wrap_content"
            android:layout_height="wrap_content"
            android:background="@drawable/shape_msg"
            android:gravity="center"
            android:orientation="horizontal"
            android:paddingLeft="30dp"
            android:paddingTop="5dp"
            android:paddingRight="30dp"
            android:paddingBottom="5dp">

            <TextView
                android:id="@+id/lob"
                android:layout_width="wrap_content"
                android:layout_height="wrap_content"
                android:layout_weight="1"
                android:onClick="lightOn"
                android:text=" 开启 "
                android:textColor="@android:color/darker_gray"
                android:textSize="30sp"/>

            <TextView
                android:layout_width="2dp"
                android:layout_height="match_parent"
                android:layout_marginLeft="5dp"
                android:layout_marginRight="5dp"
                android:background="#7fffffff"/>

            <TextView
                android:id="@+id/ldb"
                android:layout_width="wrap_content"
                android:layout_height="wrap_content"
                android:layout_weight="1"
                android:onClick="lightOff"
                android:text=" 关闭 "
                android:textColor="#fcf404"
                android:textSize="30sp"/>
```

```xml
        </LinearLayout>
    </LinearLayout>

    <LinearLayout
        android:layout_width="match_parent"
        android:layout_height="0dp"
        android:layout_weight="3"
        android:gravity="center"
        android:orientation="vertical">

        <ImageView
            android:id="@+id/feng"
            android:layout_width="@android:dimen/thumbnail_height"
            android:layout_height="@android:dimen/thumbnail_width"
            android:src="@drawable/deng"/>
    </LinearLayout>

</LinearLayout>
```

图 2-6 效果图

⑤ 在 AndroidManifest.xml 配置文件中，添加网络权限：

`<uses-permission android:name="android.permission.INTERNET"/>`

⑥ 修改 MainActivity.java。我们在 MainActivity.java 中增加四个成员变量，包括两个文本对象（将 TextView 做成按钮样式）、一个复选框对象和一个图片，代码如下。

```
private TextView lob;
    private TextView ldb;
           private CheckBox checkBox;
private ImageView feng;
```

⑦ 修改 MainActivity.java。在 onCreate() 方法中的条用的 initView() 方法中初始化【开启】按钮、【关闭】按钮和【键值发送】复选框，并将它们的状态设置为不可用，代码如下。

```
lob=findViewById(R.id.lob);
ldb=findViewById(R.id.ldb);
feng=findViewById(R.id.feng);
checkBox=findViewById(R.id.type);
```

⑧ 修改 MainActivity.java。添加【开启】按钮方法、【关闭】按钮方法，并在开启或关闭是判断【键值发送】复选框可用状态，判断【键值发送】复选框是否打钩，如果是，则以键值方式发送【开启】命令，并且键值每发送一次就会听到蜂鸣器响一次，该方式可以重复发送键值到设备，比如当设备收到【开启】命令后，如果再以键值形式继续发送【开启】命令，设备会再次执行【开启】命令。如果【键值发送】复选框没有选中，则是以变量值方式发送命令，与键值方式不同的是，该方式即使重复发送相同的变量值到设备，设备也只会执行一次，如以变量值方式发送【开启】的命令，设备开启后，再次发送【开启】命令，设备不会接收命令；当用户单击【关闭】按钮后，判断【键值发送】复选框是否被选中，如果是，则以键值方式发送【关闭】命令；如果【键值发送】复选框没有被选中，则是以变量值方式发送【关闭】命令，代码如下。

```
//灯光开启
 public void lightOn(View view) {
        String url=checkBox.isChecked()
               ?"http://192.168.0.22:8080/wziot/wzIotApi/controlSensorByKey/
d4e9f9af-1482-4be9-a096-63f7ae015a03?uuid=30015&key=1"
               :"http://192.168.0.230:8080/wziot/wzIotApi/controlSensorByVariable/
adeef6b9-c08d-46af-9af6-3d46685ad184?uuid=30015&index=1&variable=1";
        wht.setHttpURL(url);
        wht.sendControl();
  }
//灯光关闭
    public void lightOff(View view) {
        String url=checkBox.isChecked()
               ?"http://192.168.0.22:8080/wziot/wzIotApi/controlSensorByKey/
d4e9f9af-1482-4be9-a096-63f7ae015a03?uuid=30015&key=0"
               :"http://192.168.0.230:8080/wziot/wzIotApi/controlSensorByVariable/
adeef6b9-c08d-46af-9af6-3d46685ad184?uuid=30015&index=1&variable=2";
        wht.setHttpURL(url);
        wht.sendControl();
    }
```

⑨ 当用户将"开启"或"关闭"命令发送到物联网融合平台后，平台返回用户发送状态码以及发送信息，用户根据发送状态码判断加载不同的图片，代码如下。

```
private void show(String jsonResult) {
```

```
        try {
            JSONObject obj=new JSONObject(jsonResult);
            String code=obj.getString("code");
            if ("202".equals(code)) {
                if (!isLightOn) {
                    isLightOn=true;
                    lob.setTextColor(Color.parseColor("#fcf404"));
                    feng.setBackgroundResource(R.drawable.deng_on);
                    ldb.setTextColor(Color.GRAY);
                } else {
                    isLightOn=false;
                    lob.setTextColor(Color.GRAY);
                    feng.setBackgroundResource(R.drawable.deng);
                    ldb.setTextColor(Color.parseColor("#fcf404"));
                }
            }
        } catch (JSONException e) {
            e.printStackTrace();
        }
    }
```

 任务小结

请同学们根据完成情况对完成本次任务的知识、技能等要点进行小结。

任务 1 小结	
知识点掌握情况	
技能点掌握情况	

任务拓展　通风控制

 任务描述

智能楼宇的新风系统，会每隔一段时间进行室内空气的通风排气，把室内浑浊的气体排到室外，将室内 PM2.5 指数始终维持在健康范围以内，改善空气品质。

 **任务分析**

安卓移动设备控制风扇的开启和关闭，原理与控制照明设备相似，也是通过安卓移动设备端发送指令到物联网融合平台，平台将命令发送到网关，网关发送道单片机，单片机再利用继电器，来控制"开启"风扇和"关闭"风扇。

定时器的使用

在我们 Android 客户端上有时候可能有些任务不是当时就执行，而是过了一个规定的时间再执行此次任务，那么这个时候定时器就非常有用了。关键代码如下。

```
Timer timer=new Timer();
timer.schedule(new TimerTask() {
    @Override
    public void run() {
        /**
         * 要循环执行的代码
         */
    }
},1,1000);// 时间以毫秒为单位
```

Timer 是 Android 直接启动定时器的类，schedule() 方法的第一个参数是 TimerTask 类的对象，它是一个子线程，方便处理一些比较复杂耗时的功能逻辑，要实现 TimerTask 的 run() 方法，即要周期执行的一个任务；第二个参数代表从定时器初始化成功，开始启动的延迟时间，单位是秒；第三个参数代表定时器的间隔时间（执行的周期，long 类型），单位是毫秒。

任务实现

具体操作步骤如下。

① 在 Android Studio 中创建一个新项目，将应用名称设置为 demo_fs_ex，并为活动添加一个空活动。

② 为了实现隔一段时间，自动开启风扇通风，需要创建一个定时器对象，定时发送开启风扇命令，修改 MainActivity.java，加入代码如下。

```
// 创建一个定时器对象
Timer timer=new Timer();
// 创建定时器任务对象，必须实现 run() 方法，在该方法中定义用户任务
TimerTask task=new TimerTask() {
    @Override
    public void run(){
        try {
            String url=checkBox.isChecked()
                ?"http://192.168.0.22:8080/wziot/wzIotApi/controlSensorByKey/d4e9f9af-1482-4be9-a096-63f7ae015a03?uuid=30012&key=1"
                :"http://192.168.0.22:8080/wziot/wzIotApi/controlSensorByVariable/d4e9f9af-1482-4be9-a096-63f7ae015a03?uuid=30012&index=1&variable=1";
            wht.setHttpURL(url);
            wht.sendControl();
        } catch (Exception e){
            e.printStackTrace();
        }
    }
};
```

③ 在 onCreate() 方法中启动定时器，每隔 5 s 启动一次定时器，代码如下。

```
// 启动定时器
timer.schedule(task,0,5000);
```

④ 实现效果图如图 2-7 所示。

图 2-7　效果图

任务 2　智能考勤

 任务描述

人员管理是楼宇责任化管理的重中之重，在大楼出入口安装 RFID 读写设备，员工门禁卡实名办理、访客实名登记，执行严格的出入管理流程，还可配合视频监控、入侵报警等安全防范系统。

 **任务分析**

每一个员工卡中都内置了 RFID 芯片，当员工通过大楼出入口通道刷卡时，RFID 读卡器读取到员工卡中的信息并解码，然后发送至网关，网关将员工卡信息发送到物联网融合平台，Android 端的程序通过物联网融合平台接口读取 RFID 标签数据，再和已有的员工信息列表进行比对，如果存在该员工，则考勤登记，如果不存在，则提示"非法入侵"。

 知识引入

1. RFID 原理

RFID（Radio Frequency Identification）技术，又即无线射频识别，可通过无线电信号识别特定目标并读写相关数据，而无须识别系统与特定目标之间建立机械或光学接触。

RFID 读卡器（见图 2-8）是一种能阅读电子标签数据的自动识别设备。当带有 RFID 标签的员工卡进入磁场后，接收 RFID 读卡器发出的射频信号，凭借感应电流所获得的能量发送出存储在芯片中的产品信息，或者主动发送某一频率的信号，发送存储在芯片中的信息，RFID 读卡器读取信息并解码后，再送至中央信息系统进行有关数据处理。

图 2-8　RFID 读卡器照片

2. colors.xml 中的颜色定义

制作界面时我们经常会用到一些颜色，可以在 res 文件目录下的 values 里 colors.xml 文件中将要用到的颜色定义出来，示例代码如下。

```xml
<resources>
    <color name="white">#FFFFFF</color><!-- 白色 -->
    <color name="ivory">#FFFFF0</color><!--象牙色 -->
</resources>
```

3. strings.xml 中数组定义

除了在 Java 代码中定义数组，Android 还提供了在资源中定义数组，然后在 Java 代码中解析资源，从而获取数组的方法。实际开发中，推荐将数据存放在资源文件中，以实现程序的逻辑代码与数据分离，便于项目的管理，尽量减少对 Java 代码的修改。Android 规定存放数组的文件必须在 res/values 文件夹下创建，在 strings.xml 中定义。例如，定义了一个含有四个直辖市名称的字符串数组，数组名是 citys，数组元素在 <item> 标签中存放，代码如下。

```xml
<resources>
    <string-array name="citys">
        <item> 北京 </item>
        <item> 天津 </item>
        <item> 上海 </item>
        <item> 重庆 </item>
```

```
        </string-array>
</resources>
```

4. Resources 类

Android 提供了 Resources 类，通过该类提供的方法可以很方便地获取资源中的数据，如资源中定义的数组。

getResources() 方法是 ContextWrapper 类的静态方法，用于创建 Resources 对象，并且该方法必须在 Context 类及其子类中才能使用。创建一个 Resources 对象，示例代码如下。

```
Resources res=Resources.getResources();
```

Resources 类的常用方法 getStringArray(int resId) 用于获取资源索引值为 resId 的字符串类型的数组。要将创建的 citys 数组获取并存放在数组 citys 中，示例代码如下。

```
Resources res=getResources();
String[] citys=res.getStringArray(R.array.citys);
```

Resources 还提供了获取 int、boolean 等类型的数组的方法。

5. 相关的 API 方法

查询单个传感器最新数据：

请求方式：GET。

请求地址：http://192.168.0.193:8080/wziot/wzIotApi/getOneSensorData/{projectId}/{uuid}。

URL 请求参数与响应参数如图 2-9 所示。

URL请求参数		
参数	类型	描述
projectId	String	项目ID
uuid	String	传感器UUID

响应参数		
参数	类型	描述
code	int	返回状态码
msg	String	返回的消息
res	Collection of SdkSensorData	

图 2-9　URL 请求参数与响应参数

SdkSensorData 类中参数说明如图 2-10 所示。

参数	类型	描述
uuid	String	传感器UUID
passGatewayNum	String	传感器通道号
value	String	传感器当前通道的值
time	String	数据上传时间戳

图 2-10　参数说明

项目二　智能楼宇

任务实现

这里将使用安卓移动设备界面上的按钮和复选框来实现"打开"或"关闭"照明灯的功能，具体操作步骤如下。

① 在 Android Studio 中创建一个新项目，将应用名称设置为 demo_dk，并为活动添加一个空活动。

② 将项目相关图片和 RFID 图片复制到 res/drawable 文件夹下，如图 2-1 所示。

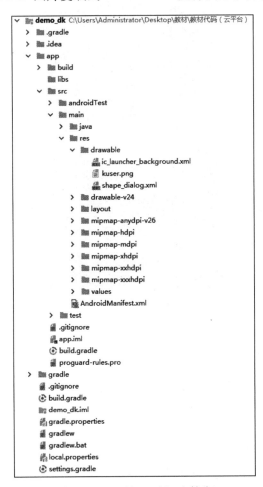

图 2-11　res/drawable 文件夹

③ 在资源文件 res/values/colors.xml 中加入字体颜色，代码如下。

```
<resources>
        <color name="colorText">#FFFFFF</color>
</resources>
```

④ 在资源文件 res/values/ strings.xml 中定义字符串数组，代码如下。

```
<resources>
```

```
    <string name="app_name">StaffAttendance</string>
    <string-array name="StaffData">
        <item>12345678</item>
        <item>23456789</item>
    </string-array>
</resources>
```

⑤ 修改 activity_main.xml，在主活动布局中添加控件，代码如下，实现效果图如图 2-12 所示。

图 2-12　效果图

```
<ImageView
    android:id="@+id/imageView"
    android:layout_width="80dp"
    android:layout_height="wrap_content"
    app:srcCompat="@drawable/kuser"/>

<LinearLayout
    android:layout_width="wrap_content"
    android:layout_height="match_parent"
    android:layout_marginLeft="20dp"
    android:orientation="vertical">

    <TextView
        android:id="@+id/uid"
        android:layout_width="wrap_content"
        android:layout_height="0dp"
        android:layout_weight="1"
        android:gravity="center_vertical"
        android:text="用户："
        android:textColor="#578DFF"
        android:textSize="24sp"/>
```

```xml
<TextView
    android:id="@+id/textView2"
    android:layout_width="wrap_content"
    android:layout_height="0dp"
    android:layout_weight="1"
    android:gravity="center"
    android:text="打卡成功！"
    android:textColor="#578DFF"
    android:textSize="24sp"/>
</LinearLayout>
```

⑥ 创建 MyAlertDialog.java 类。自定义弹出框样式，并实现值的设置，代码如下。

```java
public class MyAlertDialog extends AlertDialog {

    private Context context;

    private TextView uid;

    private Handler handler=new Handler();

    protected MyAlertDialog(Context context) {
        super(context);
        this.context=context;
    }

    @Override
    protected void onCreate(Bundle savedInstanceState) {
        super.onCreate(savedInstanceState);
        // 提前设置 Dialog 的一些样式
        Window dialogWindow=getWindow();
        dialogWindow.setGravity(Gravity.CENTER);// 设置 dialog 显示居中
        getWindow().setBackgroundDrawableResource(R.drawable.shape_dialog);
        setContentView(R.layout.dialog_user);
        setWidth();
        initView();
    }

    private void setWidth() {
        WindowManager windowManager=((Activity) context).getWindowManager();
        Display display=windowManager.getDefaultDisplay();
        WindowManager.LayoutParams lp=getWindow().getAttributes();
        lp.width=display.getWidth()*3/5;// 设置 dialog 宽度为屏幕的 1/2
        getWindow().setAttributes(lp);
    }

    private void initView() {
        uid=findViewById(R.id.uid);
    }

    public void setText(CharSequence text) {
        uid.setText(text);
```

```
    }

    @Override
    protected void onStart() {
        super.onStart();
        handler.postDelayed(new Runnable() {
            @Override
            public void run() {
                MyAlertDialog.this.dismiss();
            }
        }, 4000);
    }
}
```

⑦ 修改 MainActivity.java。在 onCreate() 方法中的 initView() 方法，初始化 MyAlertDialog 类代码如下。

```
private MyAlertDialog myAlertDialog;
private void initView() {
    myAlertDialog=new MyAlertDialog(MainActivity.this);
}
```

⑧ 设置 HTTP 请求地址，并启动定时器，每隔 2 s 对 API 接口进行一次请求，代码如下。

```
wht=new Wz_HttpTools(handler);
    wht.setHttpURL("http://192.168.0.193:8080/wziot/wzIotApi/getOneSensorData/3179a728-51c4-4fcc-9454-d7324c72187d/30013");
TimerTask task=wht.getJsonData();
timer.schedule(task, 2000, 2000);
```

⑨ 与平台连接成功后，通过 Wz_HttpTools.java 中的 getJsonData() 方法获取数据，代码如下。

```
public TimerTask getJsonData() {
    return task=new TimerTask() {
        @Override
        public void run() {
            // TODO Auto-generated method stub
            new Thread(new Runnable() {
                @Override
                public void run() {
                    String s=getConnn();
                    if (s!=null) {
                        Message message=new Message();
                        message.what=1;
                        message.obj=s;
                        handler.sendMessage(message);
                    }
                }
            }).start();
        }
    };
}
```

⑩ 将通过 http 请求获取到的 RFID 值发送到主线程的 Handler 中处理，Handler 将接收到的 RFID 值传递给自定义方法 show(String jsonResult)，代码如下。

```
Handler handler=new Handler() {

    @Override
    public void handleMessage(Message msg) {
        String jsonResult=msg.obj.toString();
        // TODO Auto-generated method stub
        // 要做的事情
        switch (msg.what) {
           case 1:
               show(jsonResult);
               break;
        }
    }
};
```

⑪ 修改 MainActivity.java。在 show(String jsonResult) 方法中，将接收到的 RFID 值显示在界面的文本框中，将资源文件中的数组初始化为字符串数组，然后遍历数组，判断是否查找到相同的 RFID 值，如果有，代表员工"打卡成功"，否则，则显示"非法入侵"，代码如下。

```
private void show(String jsonResult) {
    try {
        String code=new JSONObject(jsonResult).getString("code");
        if (!"201".equals(code))
            return;
        JSONArray obj=new JSONObject(jsonResult).getJSONArray("res");
        for (int i=0; i<obj.length(); i++) {
            JSONObject json=(JSONObject) obj.get(i);
            String rindex=json.getString("passGatewayNum");
            if ("".equals(rindex)) {
                String dtime=json.getString("time");

                if (dtime.equals(time)) {
                    return;
                }
                if (time==null) {
                    time=dtime;
                }
                final String value=json.getString("value");
                runOnUiThread(new Runnable() {
                    @Override
                    public void run() {
                        myAlertDialog.show();
                        myAlertDialog.setText(value);

                    }
                });
            }
        }
    } catch (JSONException e) {
```

```
            e.printStackTrace();
        }
    }
```

⑫ 程序运行效果如图 2-13 所示。

图 2-13　打卡成功

 任务小结

请同学们根据完成情况对完成本次任务的知识、技能等要点进行小结。

任务 2 小结	
知识点掌握情况	
技能点掌握情况	

任务拓展　打卡语音播报

 任务描述

当员工打卡成功时，语音播报"您好"，如果不存在该员工，则提示"来访用户请登记"。

 任务分析

在之前项目基础上，加入语音播报器的控制。

 **任务实现**

实现步骤如下。

① 在 Android Studio 中创建一个新项目,将应用名称设置为 demo_dk_ex,并为活动添加一个空活动。

② 在 onCreate() 方法中添加控制语音播报设备,代码如下。

```
sWht=new Wz_HttpTools(handler);
sWht.setHttpURL("http://192.168.0.193:8080/wziot/wzIotApi/controlSensor ByVariable/
3179a728-51c4-4fcc-9454-d7324c72187d?uuid=30023&index=1&variable=1");
```

③ 获取 RFID 设备值的方法和之前项目中相同,当 RFID 值和数组中的卡号相同时,修改 MainActivity.java 中的 show(String jsonResult) 方法。

```
private void show(String jsonResult) {
    try {
        String code=new JSONObject(jsonResult).getString("code");
        if (!"201".equals(code))
            return;
        JSONArray obj=new JSONObject(jsonResult).getJSONArray("res");
        for (int i=0; i<obj.length(); i++) {
            JSONObject json=(JSONObject) obj.get(i);
            String rindex=json.getString("passGatewayNum");
            if ("".equals(rindex)) {
                String dtime=json.getString("time");

                if (dtime.equals(time)) {
                    return;
                }
                if (time==null) {
                    time=dtime;
                }
                final String value=json.getString("value");
                runOnUiThread(new Runnable() {
                    @Override
                    public void run() {
                        myAlertDialog.show();
                        myAlertDialog.setText(value);
                    }
                });
                sWht.sendControl();
            }
        }
    } catch (JSONException e) {
        e.printStackTrace();
    }
}
```

任务 3　楼宇文化宣传

任务描述

LED 广告通常安装在人流量较大的商业区楼宇的外立面上，在商业区铺天盖地的广告面前，受众记忆空间的有限性和信息传播的无限性，已使注意力逐渐成为一种稀缺资源，用 LED 显示屏播放公司广告，传播速度快、形式新颖，易于被人们接受，往往会收到很好的效果。

任务分析

客户端通过唯众物联网融合平台对 LED 显示内容进行控制。

知识引入

1. LED 点阵屏的原理

LED 点阵屏通过 LED（发光二极管）组成，以灯珠亮灭来显示文字、图片、动画、视频等，是各部分组件都模块化的显示器件，通常由显示模块、控制系统及电源系统组成。

一般我们使用点阵显示汉字时多使用 16×16 的点阵宋体字库，所谓 16×16，是每一个汉字在纵、横各 16 点的区域内显示的。也就是说，可用四个 8×8 点阵组合成一个 16×16 的点阵。

2. 相关 API 方法

键值控制：

请求方式：GET。

请求地址：http:// 融合平台 IP:8080/wziot/wzIotApi/controlSensorByKey/{projectId}?uuid={uuid}&key= {key}。

URL 请求参数和响应参数如图 2-14 所示。

URL请求参数		
参数	类型	描述
projectId	String	项目ID
uuid	String	传感器UUID
key	int	键值
响应参数		
参数	类型	描述
code	int	返回状态码
msg	String	返回的消息

图 2-14　URL 请求参数和响应参数

变量值控制：

请求方式：GET。

请求地址：http:// 融合平台 IP:8080/wziot/wzIotApi/controlSensorByVariable/{projectId}?uuid=30

012&index=1&variable=1。

URL 请求参数和响应参数如图 2-15 所示。

URL请求参数		
参数	类型	描述
projectId	String	项目ID
uuid	String	传感器uuid
index	String	索引值
variable	int	变量值

响应参数		
参数	类型	描述
code	int	返回状态码
msg	String	返回的消息

图 2-15　URL 请求参数和响应参数

任务实现

① 登录唯众物联网融合平台，创建项目，在项目下添加设备，添加成功后单击【生成 API】按钮。

② 在 Android Studio 中创建一个新项目，将应用名称设置为 GZDemo，并为活动添加一个空活动。

③ 将 LED 的图片复制到 res/drawable 文件夹下，如图 2-16 所示。

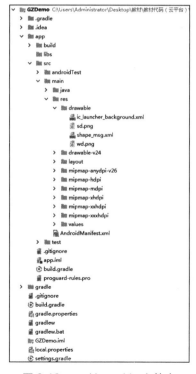

图 2-16　res/drawable 文件夹

④ 修改 activity_main.xml，在主活动布局中添加控件，代码如下，实现效果如图 2-17 所示。

```xml
<LinearLayout
    android:layout_width="match_parent"
    android:layout_height="match_parent"
    android:background="#01123D"
    android:gravity="center"
    android:orientation="vertical">

    <LinearLayout
        android:layout_width="wrap_content"
        android:layout_height="wrap_content"
        android:background="@drawable/shape_button"
        android:gravity="center"
        android:orientation="horizontal">

        <TextView
            android:id="@+id/conn"
            android:layout_width="wrap_content"
            android:layout_height="wrap_content"
            android:layout_marginLeft="20dp"
            android:layout_marginTop="10dp"
            android:layout_marginRight="20dp"
            android:layout_marginBottom="10dp"
            android:text=" 显示 "
            android:textColor="#3f71df"
            android:textSize="24sp"/>

        <TextView
            android:id="@+id/textView2"
            android:layout_width="3dp"
            android:layout_height="match_parent"
            android:background="#4d74E2"
            android:textSize="24sp"/>

        <TextView
            android:id="@+id/close"
            android:layout_width="wrap_content"
            android:layout_height="wrap_content"
            android:layout_marginLeft="20dp"
            android:layout_marginTop="10dp"
            android:layout_marginRight="20dp"
            android:layout_marginBottom="10dp"
            android:text=" 关闭 "
            android:textColor="#7f8587"
            android:textSize="24sp"/>

    </LinearLayout>

    <LinearLayout
        android:layout_width="wrap_content"
        android:layout_height="wrap_content"
        android:orientation="horizontal">
```

```xml
        <CheckBox
            android:id="@+id/cbKey"
            android:layout_width="wrap_content"
            android:layout_height="wrap_content"
            android:layout_marginTop="20dp"
            android:layout_marginBottom="20dp"
            android:layout_weight="1"
            android:text=" 键值发送 "
            android:textColor="@android:color/white"
            android:textSize="24sp"/>
    </LinearLayout>

    <LinearLayout
        android:layout_width="match_parent"
        android:layout_height="200dp"
        android:layout_marginLeft="50dp"
        android:layout_marginRight="50dp"
        android:layout_marginBottom="100dp"
        android:background="@drawable/shape_msg"
        android:gravity="center"
        android:orientation="horizontal">

        <TextView
            android:id="@+id/tvValue"
            android:layout_width="wrap_content"
            android:layout_height="wrap_content"
            android:text=" 唯众智创 "
            android:textColor="@android:color/holo_red_light"
            android:textSize="80sp"/>
    </LinearLayout>
</LinearLayout>
```

图 2-17　LED 显示

⑤ 当用户单击【显示】按钮后，判断【键值发送】复选框是否打钩，如果是，则以键值方式发送【显示】命令，否则则是以变量值方式发送命令；当用户单击【关闭】按钮时，代表更换 LED 显示内容，代码如下。

```
    // 开启
    case R.id.conn:
        System.out.println(cbKey.isChecked());
        if (cbKey.isChecked()) {
            // 设置键值 URL
            whtSend.setHttpURL("http://192.168.0.22:8080/wziot/wzIotApi/controlSensorByKey/d4e9f9af-1482-4be9-a096-63f7ae015a03?uuid=30021&key=1");
        } else {
            // 设置变量值 URL
            whtSend.setHttpURL("http://192.168.0.22:8080/wziot/wzIotApi/controlSensorByVariable/d4e9f9af-1482-4be9-a096-63f7ae015a03?uuid=30021&&index=1&variable=0");
        }
        whtSend.sendControl();
        value.setText("LED 显示: 1");
        break;
    // 关闭
    case R.id.close:
        if (cbKey.isChecked()) {
            // 设置键值 URL
            whtSend.setHttpURL("http://192.168.0.22:8080/wziot/wzIotApi/controlSensorByKey/d4e9f9af-1482-4be9-a096-63f7ae015a03?uuid=30021&key=2");
        } else {
            // 设置变量值 URL
            whtSend.setHttpURL("http://192.168.0.22:8080/wziot/wzIotApi/controlSensorByVariable/d4e9f9af-1482-4be9-a096-63f7ae015a03?uuid=30021&&index=1&variable=0");
        }
        value.setText("LED 显示: 2");
        whtSend.sendControl();
        break;
}
```

任务小结

请同学们根据完成情况对完成本次任务的知识、技能等要点进行小结。

任务 3 小结	
知识点掌握情况	
技能点掌握情况	

任务拓展： 楼宇室内舒适度显示

任务描述

在楼宇的室内，我们可以通过安装传感器来监测空间的环境状况，室内温湿度的水平严重影响人体的舒适度，因此可以在 LED 显示屏中加入温湿度的测量。

任务分析

在项目之前中加入对温湿度传感器的关注，通过接收传感器发送来的数值，判断是否在人体舒适度范围内，由此发出不同的 LED 控制指令。例如，当温度为 25～28℃，湿度为 60%～80% 时，LED 显示"室内舒适度适宜"。

任务实现

实现步骤如下。

① 登录唯众物联网融合平台，创建项目，在项目下添加设备，添加成功后单击【生成 API】按钮。

② 在 Android Studio 中创建一个新项目，将应用名称设置为 GZDemo_ex，并为活动添加一个空活动。

③ 将项目相关的图片复制到 res/drawable 文件夹下，如图 2-18 所示。

④ 修改 activity_main.xml，在主活动布局中添加控件，代码如下，实现效果如图 2-19 所示。

```xml
<LinearLayout
    android:layout_width="match_parent"
    android:layout_height="match_parent"
    android:background="#01123D"
    android:orientation="vertical">

    <LinearLayout
        android:layout_width="match_parent"
        android:layout_height="wrap_content"
        android:layout_marginLeft="30dp"
        android:layout_marginRight="30dp"
        android:layout_marginBottom="10dp"
        android:orientation="horizontal">

        <LinearLayout
            android:layout_width="match_parent"
            android:layout_height="wrap_content"
            android:layout_marginRight="5dp"
            android:layout_weight="1"
            android:background="@drawable/shape_msg"
            android:orientation="horizontal"
```

```xml
        android:paddingLeft="20dp"
        android:paddingRight="20dp">

        <ImageView
            android:id="@+id/imageView"
            android:layout_width="200dp"
            android:layout_height="wrap_content"
            android:layout_weight="1"
            tools:srcCompat="@drawable/wd"/>

        <LinearLayout
            android:layout_width="match_parent"
            android:layout_height="match_parent"
            android:layout_weight="1"
            android:orientation="vertical">

            <TextView
                android:id="@+id/tvshow"
                android:layout_width="match_parent"
                android:layout_height="0dp"
                android:layout_weight="1"
                android:gravity="center"
                android:text=" 温度 "
                android:textColor="#6092f1"
                android:textSize="30sp"/>

            <TextView
                android:id="@+id/textView"
                android:layout_width="match_parent"
                android:layout_height="0dp"
                android:layout_weight="1"
                android:gravity="center"
                android:text=" 无数据 "
                android:textSize="24sp"/>
        </LinearLayout>

    </LinearLayout>

    <LinearLayout
        android:layout_width="match_parent"
        android:layout_height="wrap_content"
        android:layout_marginLeft="5dp"
        android:layout_weight="1"
        android:background="@drawable/shape_msg"
        android:orientation="horizontal"
        android:paddingLeft="20dp"
        android:paddingRight="20dp">

        <ImageView
            android:id="@+id/imageView3"
            android:layout_width="200dp"
            android:layout_height="wrap_content"
            android:layout_weight="1"
            tools:srcCompat="@drawable/sd"/>
```

```xml
        <LinearLayout
            android:layout_width="match_parent"
            android:layout_height="match_parent"
            android:layout_weight="1"
            android:orientation="vertical">

            <TextView
                android:id="@+id/textView4"
                android:layout_width="match_parent"
                android:layout_height="0dp"
                android:layout_weight="1"
                android:gravity="center"
                android:text=" 湿度 "
                android:textColor="#6092f1"
                android:textSize="30sp"/>

            <TextView
                android:id="@+id/tvValue"
                android:layout_width="match_parent"
                android:layout_height="0dp"
                android:layout_weight="1"
                android:gravity="center"
                android:text=" 无数据 "
                android:textSize="24sp"/>
        </LinearLayout>

    </LinearLayout>
</LinearLayout>

<LinearLayout
    android:layout_width="match_parent"
    android:layout_height="300dp"
    android:layout_marginLeft="30dp"
    android:layout_marginRight="30dp"
    android:background="@drawable/shape_msg"
    android:gravity="center"
    android:orientation="horizontal">

    <TextView
        android:id="@+id/msg"
        android:layout_width="wrap_content"
        android:layout_height="wrap_content"
        android:text=" 室内温度适宜 "
        android:textColor="@android:color/holo_red_light"
        android:textSize="70sp"/>
</LinearLayout>

</LinearLayout>
```

物联网移动应用开发

图 2-18　res/drawable 文件夹　　　　图 2-19　室内温度

⑤ 用户启动定时器。注：定时器获取数据间隔最小时间为 2s，开始接收温湿度传感器数据，代码如下。

```
whtData.setHttpURL("http://192.168.0.22:8080/wziot/wzIotApi/getOneSensorData/d4e9f9af-1482-4be9-a096-63f7ae015a03/30008");
TimerTask task=whtData.getJsonData();
timer.schedule(task, 2000, 2000);
```

⑥ 与平台连接成功后，通过 Wz_HttpTools.java 中的 getJsonData() 方法获取数据，代码如下。

```
public TimerTask getJsonData() {
    return task=new TimerTask() {
        @Override
        public void run() {
            // TODO Auto-generated method stub
            new Thread(new Runnable() {
                @Override
                public void run() {
                    String s=getConnn();
                    if (s!=null) {
```

56

```
                    Message message=new Message();
                    message.what=1;
                    message.obj=s;
                    handler.sendMessage(message);
                }
            }
        }).start();
    }
};
```

⑦ 当获取到传感器数据时，将传感器值发送到主线程的 Handler 中处理，标识码为 1，修改 MainActivity.java 的代码，分别处理消息，代码如下。

```
Handler handler=new Handler() {
    @Override
    public void handleMessage(Message msg) {
        // TODO Auto-generated method stub

        switch (msg.what) {
            case 1:
                show(msg.toString());
                break;

        }
    }
};
```

⑧ 在 MainActivity.java 中加入 show(String jsonResult) 方法，当温度在人体舒适度范围内时，发出 LED 控制指令，此时将在 LED 屏上看到"温度适合"，代码如下。

```
public void show(String jsonResult){
    try{
        String s="";
        JSONObject jres=new JSONObject(jsonResult);// 用 JSONObject 进行解析
        String msg=jres.getString("msg");// 获取返回信息
        s+=" 状态: "+msg+"\n";
        JSONArray obj=jres.getJSONArray("res");
        int wenValue=0;
        int shiValue=0;
        for(int i=0;i<obj.length();i++){
            JSONObject json=(JSONObject)obj.get(i);
            String index=json.getString("index");
            String uuid=json.getString("uuid");
            String value=json.getString("value");
            s+=" 通道号: "+index+",UUID: "+uuid+", 值:"+value+"\n";
            if(index.equals("1")){
                wenValue=Integer.parseInt(value);
            }else{
                shiValue=Integer.parseInt(value);
            }
        }
```

```
        // 温度为 25～28℃，湿度为 60%～80%
        if((wenValue>=25&&wenValue<=28)&&(shiValue>=60&&shiValue<=80)){
            whtSend.setHttpURL("http://192.168.0.22:8080/wziot/wzIotApi/
controlSensorByKey/d4e9f9af-1482-4be9-a096-63f7ae015a03?uuid=30021&key=2");
            whtSend.sendControl();
        }
    }catch(JSONException e){
        e.printStackTrace();
    }
}
```

项目三 智慧消防

项目概述

基于物联网构建的智慧消防,能对消防设施运行数据、各类信息进行获取与分析,实现对隐患监测与自动预警,以保障各单位的消防安全。配合大数据云计算平台、火警智能研判、单位智能监管、救援智能指挥等专业应用,可实现城市消防的智能化。

智慧消防主要包括以下内容:

智能消防能可燃气体传感器的数值,当可燃气体浓度超过阈值时,在界面显示报警图标,语音报警并打开风扇;当浓度恢复到阈值以内时,语音提示并关闭风扇;实现智慧消防联动控制,并将警告信息发送到客户端的手机上。

工作任务

- 任务1:可燃气体数据采集。
- 任务2:智慧消防联动控制。
- 任务3:智慧消防远程联动。

学习目标

- 培养学生自主探究和解决问题的能力。
- 培养学生严谨的逻辑思维能力。

任务 1　可燃气体数据采集

任务描述

随着现代化城市的发展，天然气等易燃易爆气体在日常工作生活中的应用也越来越多。这些易燃易爆的气体在生产、运输、使用等过程中，有可能发生泄漏，造成中毒、火灾、爆炸等安全事故，严重危害人民的生命和财产的安全。在使用这些可燃气体的场合，若安装可燃气体传感器以及相应配套的联动装置，采取相应的保护措施和报警给监控部门，就能及时提醒人们采取营救措施，将危险降低，从而减少人员伤亡和财产的损失。

本次任务是利用气体传感器技术，将检测到的可燃气体浓度和标准值进行比较，当高过一定浓度值时进行相应的处理。

任务分析

通过 API 接口 http:// 服务 IP 地址：端口 /wziot/wzIotApi/getOneSensorData/{projectId}/{uuid}，采集可燃气体传感器的数值。当可燃气体浓度超过阈值时，在界面显示报警图标。

知识引入

API（Application Programming Interface，应用程序编程接口）是一些预先定义的函数，目的是提供应用程序与开发人员基于某软件或硬件得以访问一组例程的能力，而又无须访问源码，或理解内部工作机制的细节。

唯众物联网融合平台，读取物联网设备硬件数据与控制物联网设备硬件全部采取 API 接口方式。

查看项目当前状态如图 3-1 所示。

请求方式：GET。

请求地址：http://192.168.0.193:8080/wziot/wzIotApi/getProjectState/{projectId}。

URL请求参数		
参数	类型	描述
projectId	String	项目ID
响应参数		
参数	类型	描述
code	int	返回状态码
msg	String	返回的消息
res	Collection of SdkProjectData	

图 3-1　当前状态

SdkProjectData 类中参数说明如图 3-2 所示。

参数	类型	描述
projrctId	String	项目ID
projectName	String	项目名称
isCreate	Boolean	是否已经生成
isStart	Boolean	是否已经启动

图 3-2　参数说明

查看项目是否生成：

请求方式：GET。

请求地址：http://192.168.0.193:8080/wziot/wzIotApi/isCreateOfProject/{projectId}。

URL 请求参数和响应参数如图 3-3 所示。

URL请求参数

参数	类型	描述
projectId	String	项目ID

响应参数

参数	类型	描述
code	int	返回状态码
msg	String	返回的消息

图 3-3　URL 请求参数和响应参数

任务实现

这里将使用安卓移动设备界面上的按钮和复选框来实现“打开”或“关闭”消防的功能，具体操作步骤如下。(实例项目：源代码 \03.1\WZLinkGas)

① 登录唯众物联网融合平台，创建项目，在项目下添加设备，添加成功后单击【生成 API】按钮。

② 在 Android Studio 中创建一个新项目，将应用名称设置为 SFireProtect，并为活动添加一个空活动。

③ 将项目相关图片和 RFID 图片复制到 res/drawable 文件夹下，如图 3-4 所示。

物联网移动应用开发

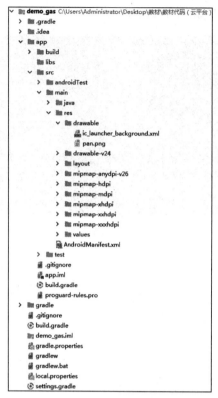

图 3-4 res/drawable 文件夹

④ 切换到【Android】视图，修改 AndroidManifest.xml，添加权限，代码如下。

```
<uses-permission android:name="android.permission.INTERNET"/>
```

⑤ 在资源文件 res/values/colors.xml 中加入字体颜色，代码如下。

```
<resources>
    <color name="colorPrimary">#008577</color>
    <color name="colorPrimaryDark">#00574B</color>
    <color name="colorAccent">#D81B60</color>
</resources>
```

⑥ 修改 activity_main.xml，在主活动布局中添加控件，代码如下，实现效果图如图 3-5 所示。

```
<LinearLayout
    android:layout_width="match_parent"
    android:layout_height="match_parent"
    android:background="#01123D"
    android:gravity="center"
    android:orientation="vertical">

    <LinearLayout
        android:layout_width="match_parent"
        android:layout_height="0dp"
        android:layout_weight="2"
```

```xml
            android:gravity="center"
            android:orientation="vertical">

            <LinearLayout
                android:layout_width="wrap_content"
                android:layout_height="wrap_content"
                android:background="@drawable/shape_msg"
                android:gravity="center"
                android:orientation="horizontal"
                android:paddingLeft="30dp"
                android:paddingTop="5dp"
                android:paddingRight="30dp"
                android:paddingBottom="5dp">

                <TextView
                    android:layout_width="wrap_content"
                    android:layout_height="wrap_content"
                    android:layout_weight="1"
                    android:text="可燃气浓度"
                    android:textColor="@android:color/white"
                    android:textSize="30sp"/>

            </LinearLayout>
        </LinearLayout>

        <LinearLayout
            android:layout_width="match_parent"
            android:layout_height="0dp"
            android:layout_weight="3"
            android:gravity="center"
            android:orientation="vertical">

            <ImageView
                android:id="@+id/imageView"
                android:layout_width="450dp"
                android:layout_height="wrap_content"
                app:srcCompat="@drawable/pan"/>

            <TextView
                android:id="@+id/svalue"
                android:layout_width="wrap_content"
                android:layout_height="wrap_content"
                android:layout_marginTop="-240dp"
                android:text="0%"
                android:textColor="#3c73e8"
                android:textSize="100sp"/>
        </LinearLayout>

</LinearLayout>
```

图 3-5　效果图

⑦ 修改 MainActivity.java。我们在 MainActivity.java 中定义并初始化成员变量，代码如下。

```
private TextView svalue;
  private void initView() {
        svalue=findViewById(R.id.svalue);
    }
```

⑧ 修改 MainActivity.java。在 onCreate() 方法中的设置请求 API 的 URL 地址，并启动定时器。

```
Wz_HttpTools wht=new Wz_HttpTools(handler);
  wht.setHttpURL("http://192.168.0.22:8080/wziot/wzIotApi/getOneSensorData/
d4e9f9af-1482-4be9-a096-63f7ae015a03/30010");
        TimerTask task=wht.getJsonData();
        timer.schedule(task, 2000, 2000);
```

⑨ 修改 MainActivity.java。在 onClick() 方法中，当网关登录成功后，修改【开启】按钮、【关闭】按钮和【键值发送】复选框为可用状态，代码如下。

```
//登录
case R.id.login:
    wzTools.LoginGateWay(ip, sn);
    connbtn.setEnabled(true);
    closebtn.setEnabled(true);
    cbKey.setEnabled(true);
    break;
```

⑩ 与平台连接成功后，通过 Wz_HttpTools.java 中的 getJsonData() 方法获取数据，代码如下。

```java
public TimerTask getJsonData() {
    return task=new TimerTask() {
        @Override
        public void run() {
            // TODO Auto-generated method stub
            new Thread(new Runnable() {
                @Override
                public void run() {
                    String s=getConnn();
                    if (s!=null) {
                        Message message=new Message();
                        message.what=1;
                        message.obj=s;
                        handler.sendMessage(message);
                    }
                }
            }).start();
        }
    };
}
private String getConnn() {
    HttpURLConnection connection=null;
    BufferedReader reader=null;
    // 第一步建立 httpurlconnection 实例
    try {
        URL url=new URL(httpURL);
        connection=(HttpURLConnection) url.openConnection();
        // 自由定制
        connection.setRequestMethod("GET");
        connection.setConnectTimeout(80000);
        connection.setReadTimeout(8000);

        // 使用 getInputStream 获取服务器返回的输入流
        InputStream stresm=connection.getInputStream();
        // 对返回的输入流进行读取
        reader=new BufferedReader(new InputStreamReader(stresm));
        StringBuilder response=new StringBuilder();
        String line;
        while ((line=reader.readLine())!=null) {
            response.append(line);
        }

        return response.toString();
    } catch (IOException e) {
        e.printStackTrace();
    } finally {
        if (reader!=null) {
            try {
                reader.close();
            } catch (IOException e) {
                e.printStackTrace();
            }
        }
        if (connection!=null) {
```

```
                connection.disconnect();
            }
        }
        return null;
    }
```

⑪ 修改 MainActivity.java。在 show(String jsonResult) 方法中，将接收到的可燃气值显示在界面的文本框中。

```
public void show(String jsonResult, String index) {
    try {
        JSONArray obj=new JSONObject(jsonResult).getJSONArray("res");
        for (int i=0; i<obj.length(); i++) {
            JSONObject json=(JSONObject) obj.get(i);
            String rindex=json.getString("passGatewayNum");
            if (index.equals(rindex)) {
                String value=json.getString("value");
                svalue.setText(value+"%");
            }
        }
    } catch (JSONException e) {
        e.printStackTrace();
    }
}
```

⑫ 运行效果如图 3-6 所示。

图 3-6　效果图

项目三 智慧消防

任务小结

请同学们根据完成情况对完成本次任务的知识、技能等要点进行小结。

任务 1 小结	
知识点掌握情况	
技能点掌握情况	

任务拓展　可燃气语音预警

任务描述

智慧消防对隐患监测与自动预警，不仅可在平台上切换图标来警示监管人员，还应该有语音提示预警。

任务分析

安卓移动设备通过唯众物联网融合平台可控制报警铃声的开启和关闭，原理与控制照明设备相似，也是通过安卓移动设备端发送指令到物联网融合平台，平台再发送指令到网关，网关发送指令到单片机，单片机再利用继电器，来控制"关闭"警铃和"打开"警铃。

任务实现

具体操作步骤如下。

① 登录唯众物联网融合平台，创建项目，在项目下添加设备，添加成功后单击【生成 API】按钮。

② 在 Android Studio 中创建一个新项目，将应用名称设置为 demo_gas_ex，并为活动添加一个空活动。

③ 将项目相关的图片复制到 res/drawable 文件夹下，如图 3-7 所示。

④ 修改 activity_main.xml，在主活动布局中添加控件，代码如下，实现效果图如图 3-8 所示。

```
<LinearLayout
    android:layout_width="match_parent"
    android:layout_height="match_parent"
    android:background="#01123D"
    android:gravity="center"
    android:orientation="vertical">

    <LinearLayout
        android:layout_width="match_parent"
        android:layout_height="0dp"
```

```xml
            android:layout_weight="2"
            android:gravity="center"
            android:orientation="vertical">

            <LinearLayout
                android:layout_width="wrap_content"
                android:layout_height="wrap_content"
                android:background="@drawable/shape_msg"
                android:gravity="center"
                android:orientation="horizontal"
                android:paddingLeft="30dp"
                android:paddingTop="5dp"
                android:paddingRight="30dp"
                android:paddingBottom="5dp">

                <TextView
                    android:layout_width="wrap_content"
                    android:layout_height="wrap_content"
                    android:layout_weight="1"
                    android:text=" 可燃气浓度 "
                    android:textColor="@android:color/white"
                    android:textSize="30sp" />

            </LinearLayout>
        </LinearLayout>

        <LinearLayout
            android:layout_width="match_parent"
            android:layout_height="0dp"
            android:layout_weight="3"
            android:gravity="center"
            android:orientation="vertical">

            <ImageView
                android:id="@+id/imageView"
                android:layout_width="450dp"
                android:layout_height="wrap_content"
                app:srcCompat="@drawable/pan" />

            <TextView
                android:id="@+id/svalue"
                android:layout_width="wrap_content"
                android:layout_height="wrap_content"
                android:layout_marginTop="-240dp"
                android:text="0%"
                android:textColor="#3c73e8"
                android:textSize="100sp" />
        </LinearLayout>

    </LinearLayout>
```

项目三 智慧消防

图 3-7 res/drawable 文件夹　　　图 3-8 效果图

⑤ 为了实现隔一段时间，开启可燃气监控报警，需要对监控的可燃气体的浓度数值进行判断，当浓度大于 20% 时开始报警，等于或低于 20% 时显示正常。30023 语音播放。

修改 MainActivity.java，加入代码如下。

```java
        public void show(String jsonResult, String index) {
            try{
                JSONArray obj=new JSONObject(jsonResult).getJSONArray("res");
                for(int i=0; i<obj.length(); i++) {
                    JSONObject json=(JSONObject) obj.get(i);
                    String rindex=json.getString("passGatewayNum");
                    if(index.equals(rindex)) {
                        String value=json.getString("value");
                        svalue.setText(value+"%");
                        if(Double.parseDouble(value)>20) {
                            if(startTime==null) {
                                startTime=System.currentTimeMillis();
                                sWht.sendControl();
                            } else {
                                if(System.currentTimeMillis()-startTime>10*1000) {
                                    startTime=System.currentTimeMillis();
                                    sWht.sendControl();
                                }
```

```
                    }
                }
            }
        }
    } catch (JSONException e) {
        e.printStackTrace();
    }
}
```

任务 2　智慧消防联动控制

任务描述

在使用这些可燃气体的场合,若安装可燃气体传感器监控,再配备相应的联动装置,采取相应的保护措施给监控部门提供报警,就能及时提醒人们采取营救措施,可将危险降低,从而减少人员伤亡和财产的损失。

本次任务是利用气体传感器技术,实时的检测可燃气体浓度和标准值进行比较,当高过一定浓度值时进行相应的,进行相应的联动处理。

任务分析

通过可燃气体浓度数值的采集,实现当浓度超过阈值时,语音报警并打开风扇;当浓度恢复到阈值以内,语音提示并关闭风扇。

知识引入

联动控制的逻辑结构如图 3-9 所示。

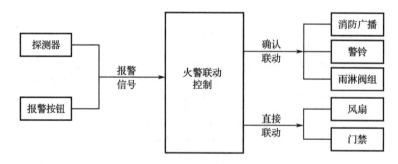

图 3-9　联动控制的逻辑结构

任务实现

这里将使用安卓移动设备接收可燃气体传感器数据值,并当值到达一定程度后,开启风扇、开启报警等消防的功能,具体操作步骤如下。

① 登录唯众物联网融合平台,创建项目,在项目下添加设备,添加成功后单击【生成 API】按钮。

② 在 Android Studio 中创建一个新项目,将应用名称设置为 demo_gas_ex_ex,并为项目添加一个空活动。

③ 将项目相关图片复制到 res/drawable 文件夹下,如图 3-10 所示。

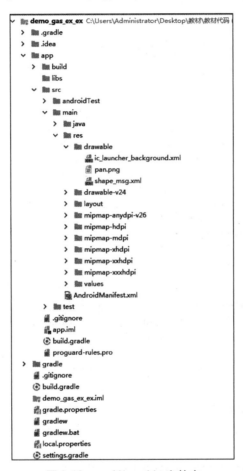

图 3-10　res/drawable 文件夹

④ 切换到【Android】视图,修改 AndroidManifest.xml,添加权限,代码如下。

`<uses-permission android:name="android.permission.INTERNET"/>`

⑤ 在资源文件 res/values/colors.xml 中加入字体颜色,代码如下。

```
<resources>
    <color name="colorPrimary">#008577</color>
    <color name="colorPrimaryDark">#00574B</color>
    <color name="colorAccent">#D81B60</color>
</resources>
```

⑥ 修改 activity_main.xml，在主活动布局中添加控件，代码如下，实现效果图如图 3-11 所示。

```xml
<LinearLayout
    android:layout_width="match_parent"
    android:layout_height="match_parent"
    android:background="#01123D"
    android:gravity="center"
    android:orientation="vertical">

    <LinearLayout
        android:layout_width="match_parent"
        android:layout_height="0dp"
        android:layout_weight="2"
        android:gravity="center"
        android:orientation="vertical">

        <LinearLayout
            android:layout_width="wrap_content"
            android:layout_height="wrap_content"
            android:background="@drawable/shape_msg"
            android:gravity="center"
            android:orientation="horizontal"
            android:paddingLeft="30dp"
            android:paddingTop="5dp"
            android:paddingRight="30dp"
            android:paddingBottom="5dp">

            <TextView
                android:layout_width="wrap_content"
                android:layout_height="wrap_content"
                android:layout_weight="1"
                android:text=" 可燃气浓度 "
                android:textColor="@android:color/white"
                android:textSize="30sp" />

        </LinearLayout>
    </LinearLayout>

    <LinearLayout
        android:layout_width="match_parent"
        android:layout_height="0dp"
        android:layout_weight="3"
        android:gravity="center"
        android:orientation="vertical">

        <ImageView
            android:id="@+id/imageView"
            android:layout_width="450dp"
            android:layout_height="wrap_content"
            app:srcCompat="@drawable/pan" />

        <TextView
            android:id="@+id/svalue"
```

```
                android:layout_width="wrap_content"
                android:layout_height="wrap_content"
                android:layout_marginTop="-240dp"
                android:text="0%"
                android:textColor="#3c73e8"
                android:textSize="100sp" />
        </LinearLayout>

</LinearLayout>
```

图 3-11 效果图

⑦ 修改 MainActivity.java。我们在 MainActivity.java 中定义成员变量,代码如下。

```
    private Wz_HttpTools wht;
    private Wz_HttpTools sWht;
    private Wz_HttpTools lWht;
        private Wz_HttpTools dWht;
    private TextView svalue;
    private Long startTime;
        private boolean hasLight=false;
```

⑧ 修改 MainActivity.java。在 onCreate() 方法中的设置请求 API 的 URL 地址,并启动定时器,代码如下。

```
wht=new Wz_HttpTools(handler);
        wht.setHttpURL("http://192.168.0.22:8080/wziot/wzIotApi/
getOneSensorData/d4e9f9af-1482-4be9-a096-63f7ae015a03/30010");
        sWht=new Wz_HttpTools(handler);
sWht.setHttpURL("http://192.168.0.22:8080/wziot/wzIotApi/
```

```
controlSensorByVariable/d4e9f9af-1482-4be9-a096-63f7ae015a03?uuid=30023&index=1
&variable=1");
        lWht=new Wz_HttpTools(handler);
        lWht.setHttpURL("http://192.168.0.22:8080/wziot/wzIotApi/
controlSensorByVariable/d4e9f9af-1482-4be9-a096-63f7ae015a03?uuid=30012&index=1
&variable=1");
        dWht=new Wz_HttpTools(handler);
        dWht.setHttpURL("http://192.168.0.22:8080/wziot/wzIotApi/
controlSensorByVariable/d4e9f9af-1482-4be9-a096-63f7ae015a03?uuid=30012&index=1
&variable=2");
        TimerTask task=wht.getJsonData();
        timer.schedule(task, 2000, 2000);
```

⑨ 与平台连接成功后，通过 Wz_HttpTools.java 中的 getJsonData() 方法获取数据，代码如下。

```
public TimerTask getJsonData() {
    return task=new TimerTask() {
        @Override
        public void run() {
            // TODO Auto-generated method stub
            new Thread(new Runnable() {
                @Override
                public void run() {
                    String s=getConnn();
                    if (s!=null) {
                        Message message=new Message();
                        message.what=1;
                        message.obj=s;
                        handler.sendMessage(message);
                    }
                }
            }).start();
        }
    };
}
private String getConnn() {
    HttpURLConnection connection=null;
    BufferedReader reader=null;
    // 第一步建立 httpurlconnection 实例
    try {
        URL url=new URL(httpURL);
        connection=(HttpURLConnection) url.openConnection();
        // 自由定制
        connection.setRequestMethod("GET");
        connection.setConnectTimeout(80000);
        connection.setReadTimeout(8000);

        // 使用 getInputStream 获取服务器返回的输入流
        InputStream stresm=connection.getInputStream();
        // 对返回的输入流进行读取
        reader=new BufferedReader(new InputStreamReader(stresm));
        StringBuilder response=new StringBuilder();
        String line;
```

```
                while ((line=reader.readLine())!=null) {
                    response.append(line);
                }
                return response.toString();
            } catch (IOException e) {
                e.printStackTrace();
            } finally {
                if (reader!=null) {
                    try {
                        reader.close();
                    } catch (IOException e) {
                        e.printStackTrace();
                    }
                }
                if (connection!=null) {
                    connection.disconnect();
                }
            }
            return null;
        }
```

⑩ 修改 MainActivity.java。在 show(String jsonResult) 方法中，将接收到的可燃气值显示在界面的文本框中，并判断可燃气体值是否超标，超标后开启风扇和语音报警，在可燃气体值降到安全值后，关闭风扇，语音播报安全信息，代码如下。

```
public void show(String jsonResult, String index) {
    try {
        String code=new JSONObject(jsonResult).getString("code");
        if (!"201".equals(code))
            return;
        JSONArray obj=new JSONObject(jsonResult).getJSONArray("res");
        for (int i=0; i<obj.length(); i++) {
            JSONObject json=(JSONObject) obj.get(i);
            String rindex=json.getString("passGatewayNum");
            if (index.equals(rindex)) {
                String value=json.getString("value");
                svalue.setText(value+"%");

                if (Double.parseDouble(value)>10) {
                    if (startTime==null) {
                        startTime=System.currentTimeMillis();
                        sWht.sendControl();
                    } else {
                        if (System.currentTimeMillis()-startTime>10*1000) {
                            startTime=System.currentTimeMillis();
                            sWht.sendControl();
                        }
                    }
                    if (!hasLight) {
                        hasLight=true;
                        lWht.sendControl();
```

```
                    }
                } else {
                    if (!hasLight) {
                        hasLight=false;
                        dWht.sendControl();
                    }
                }
            }
        } catch (JSONException e) {
            e.printStackTrace();
        }
    }
```

 任务小结

请同学们根据完成情况对完成本次任务的知识、技能等要点进行小结。

任务 2 小结	
知识点掌握情况	
技能点掌握情况	

任务拓展　设置阈值 1

 **任务描述**

可燃气体种类很多，我们前面的设置浓度阈值是 10，但现实生活中不同的气体的阈值是不同的，需要一个人机交互的界面来让用户设置合适的阈值。

 任务分析

安卓移动设备可控制阈值大小，增加按钮来进行阈值的设置和修改。

任务实现

具体操作步骤如下：

① 登录唯众物联网融合平台，创建项目，在项目下添加设备，添加成功后单击【生成 API】按钮。

② 在 Android Studio 中创建一个新项目，将应用名称设置为 demo_gas_edit，并为活动添加一个空活动。

③ 创建 dialog_set.xml，用于设置阈值弹出页面，代码如下。

```xml
<EditText
    android:id="@+id/wnum"
    android:layout_width="200dp"
    android:layout_height="wrap_content"
    android:hint="请输入警戒值"
    android:inputType="number"
    android:paddingLeft="10dp"
    android:paddingRight="10dp"
    android:textColor="#000000"/>

<Button
    android:id="@+id/done"
    android:layout_width="wrap_content"
    android:layout_height="wrap_content"
    android:text="设置"/>
```

④ 修改 activity_main.xml，在主活动布局中添加控件，代码如下，实现效果图如图 3-12 所示。

```xml
<android.support.v7.widget.Toolbar
    android:layout_width="match_parent"
    android:layout_height="wrap_content"
    android:background="?attr/colorPrimary"
    android:minHeight="?attr/actionBarSize"
    android:paddingLeft="20dp"
    android:paddingRight="20dp"
    android:theme="?attr/actionBarTheme">

    <TextView
        android:layout_width="wrap_content"
        android:layout_height="wrap_content"
        android:layout_gravity="center"
        android:text="可燃气"
        android:textSize="24sp"
        android:textStyle="bold"/>

    <ImageButton
        android:layout_width="30dp"
        android:layout_height="30dp"
        android:layout_gravity="right"
        android:background="@drawable/cl"
        android:onClick="toSet"/>
</android.support.v7.widget.Toolbar>

<LinearLayout
    android:layout_width="match_parent"
    android:layout_height="match_parent"
    android:gravity="center"
    android:orientation="vertical">

    <LinearLayout
        android:layout_width="match_parent"
        android:layout_height="0dp"
        android:layout_weight="2"
```

```xml
            android:gravity="center"
            android:orientation="vertical">

            <LinearLayout
                android:layout_width="wrap_content"
                android:layout_height="wrap_content"
                android:background="@drawable/shape_msg"
                android:gravity="center"
                android:orientation="horizontal"
                android:paddingLeft="30dp"
                android:paddingTop="5dp"
                android:paddingRight="30dp"
                android:paddingBottom="5dp">

                <TextView
                    android:layout_width="wrap_content"
                    android:layout_height="wrap_content"
                    android:layout_weight="1"
                    android:text=" 可燃气浓度 "
                    android:textColor="@android:color/white"
                    android:textSize="30sp"/>

            </LinearLayout>
        </LinearLayout>

        <LinearLayout
            android:layout_width="match_parent"
            android:layout_height="0dp"
            android:layout_weight="3"
            android:gravity="center"
            android:orientation="vertical">

            <ImageView
                android:layout_width="450dp"
                android:layout_height="wrap_content"
                app:srcCompat="@drawable/pan"/>

            <TextView
                android:id="@+id/svalue"
                android:layout_width="wrap_content"
                android:layout_height="wrap_content"
                android:layout_marginTop="-240dp"
                android:text="0%"
                android:textColor="#578DFF"
                android:textSize="100sp"/>
        </LinearLayout>

    </LinearLayout>
```

图 3-12　效果图

⑤ 添加 MyAlertDialog.java 类，设置阈值弹出窗体样式，代码如下。

```java
private Context context;

    private TextView wnum;
    private Button done;

    private StringBuilder flag;

    protected MyAlertDialog(Context context, StringBuilder flag) {
        super(context);
        this.context=context;
        this.flag=flag;
    }

    @Override
    protected void onCreate(Bundle savedInstanceState) {
        super.onCreate(savedInstanceState);
        //提前设置Dialog的一些样式
        Window dialogWindow=getWindow();
        dialogWindow.setGravity(Gravity.CENTER);//设置dialog显示居中
        getWindow().setBackgroundDrawableResource(R.drawable.shape_dialog);
        setContentView(R.layout.dialog_set);
        //显示键盘
        getWindow().clearFlags(WindowManager.LayoutParams.FLAG_ALT_FOCUSABLE_IM);
        setWidth();
        initView();
    }
```

```
    private void setWidth() {
        WindowManager windowManager=((Activity) context).getWindowManager();
        Display display=windowManager.getDefaultDisplay();
        WindowManager.LayoutParams lp=getWindow().getAttributes();
        lp.width=display.getWidth()*3/5;// 设置dialog宽度为屏幕的2/3
        getWindow().setAttributes(lp);
    }

    private void initView() {
        wnum=findViewById(R.id.wnum);
        done=findViewById(R.id.done);
        done.setOnClickListener(new View.OnClickListener() {
            @Override
            public void onClick(View v) {
                flag.replace(0, flag.length(), wnum.getText().toString());
                MyAlertDialog.this.dismiss();
                Toast toast=Toast.makeText(context, "设置成功", Toast.LENGTH_SHORT);
                toast.setGravity(Gravity.CENTER, 0, 0);
                toast.show();
            }
        });
    }
```

⑥ 修改 MainActivity.java，代码如下。

```
private Wz_HttpTools wht;
    private Wz_HttpTools sWht;
    private Wz_HttpTools lWht;
    private Wz_HttpTools dWht;

    private TextView svalue;
    private MyAlertDialog myAlertDialog;

    private Long startTime;
    private boolean hasLight=false;
    private double cct=30;
    private StringBuilder flag=new StringBuilder("");

    private final Timer timer=new Timer();
    Handler handler=new Handler() {
        @Override
        public void handleMessage(Message msg) {
            String jsonResult=msg.obj.toString();
            // TODO Auto-generated method stub
            switch (msg.what) {
                case 1:
                    show(jsonResult, "1");
                    break;
            }
            super.handleMessage(msg);
        }
```

```java
    };


    @Override
    protected void onCreate(Bundle savedInstanceState) {
        super.onCreate(savedInstanceState);
        setContentView(R.layout.activity_main);
        initView();
        wht=new Wz_HttpTools(handler);
        wht.setHttpURL("http://192.168.0.22:8080/wziot/wzIotApi/getOneSensorData/d4e9f9af-1482-4be9-a096-63f7ae015a03/30010");
        sWht=new Wz_HttpTools(handler);
        sWht.setHttpURL("http://192.168.0.22:8080/wziot/wzIotApi/controlSensorByVariable/d4e9f9af-1482-4be9-a096-63f7ae015a03?uuid=30023&index=1&variable=1");
        lWht=new Wz_HttpTools(handler);
        lWht.setHttpURL("http://192.168.0.22:8080/wziot/wzIotApi/controlSensorByVariable/d4e9f9af-1482-4be9-a096-63f7ae015a03?uuid=30012&index=1&variable=1");
        dWht=new Wz_HttpTools(handler);
        dWht.setHttpURL("http://192.168.0.22:8080/wziot/wzIotApi/controlSensorByVariable/d4e9f9af-1482-4be9-a096-63f7ae015a03?uuid=30012&index=1&variable=2");
        TimerTask task=wht.getJsonData();
        timer.schedule(task, 2000, 2000);
    }

    private void initView() {
        svalue=findViewById(R.id.svalue);
        myAlertDialog=new MyAlertDialog(MainActivity.this, flag);
    }

    public void show(String jsonResult, String index) {
        try {
            String code=new JSONObject(jsonResult).getString("code");
            if (!"201".equals(code))
                return;
            JSONArray obj=new JSONObject(jsonResult).getJSONArray("res");
            for (int i=0; i<obj.length(); i++) {
                JSONObject json=(JSONObject) obj.get(i);
                String rindex=json.getString("passGatewayNum");
                if (index.equals(rindex)) {
                    String value=json.getString("value");
                    svalue.setText(value+"%");
                    if (!"".equals(flag.toString()))
                        cct=Integer.parseInt(flag.toString());
                    if (Double.parseDouble(value) > cct) {
                        if (startTime==null) {
                            startTime=System.currentTimeMillis();
                            sWht.sendControl();
                        } else {
                            if (System.currentTimeMillis()-startTime>10*1000) {
                                startTime=System.currentTimeMillis();
                                sWht.sendControl();
```

```
                    }
                }
                if (!hasLight) {
                    hasLight=true;
                    lWht.sendControl();
                }
            } else {
                if (hasLight) {
                    System.out.println("!!!");
                    hasLight=false;
                    dWht.sendControl();
                }
            }
        }
    } catch (JSONException e) {
        e.printStackTrace();
    }

}

public void toSet(View view) {
    myAlertDialog.show();
}
```

任务 3　智慧消防远程联动

任务描述

本次任务是通过可燃气体浓度数值的采集，实时地检测可燃气体浓度，并和标准值进行比较，当高过一定浓度值时进行相应的操作，将警告信息发送到客户端的手机上。

任务分析

通过可燃气体浓度数值的采集，实现当浓度超过阈值时，将警告信息发送到客户端的手机上。

知识引入

socket 通信原理：

在同一台计算机，进程之间可以通信。如果是不同的计算机呢？网络上不同的计算机，也可以通信，此时需要使用网络套接字（socket）。socket 就是在不同计算机之间进行通信的一个抽象。其工作于 TCP/IP 协议中应用层和传输层之间。

TCP/IP（Transmission Control Protocol/Internet Protocol）即传输控制协议／网际协议，是一个工业标准的协议集，它是为广域网（WAN）设计的。

UDP（User Data Protocol，用户数据报协议）是与 TCP 相对应的协议。它是属于 TCP/IP 协议族中的一种。

TCP 与 UDP 如图 3-13 所示。

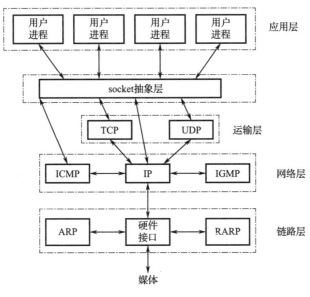

图 3-13　TCP 与 UDP

socket 是应用层与 TCP/IP 协议族通信的中间软件抽象层，它是一组接口。在设计模式中，socket 其实就是一个门面模式，它把复杂的 TCP/IP 协议族隐藏在 Socket 接口后面，对用户来说，一组简单的接口就是全部，让 socket 去组织数据，以符合指定的协议，如图 3-14 所示。

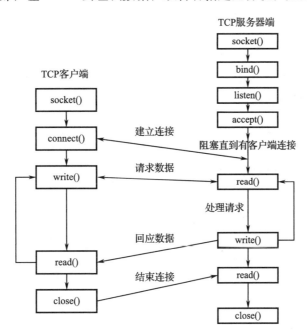

图 3-14　socket 通信

服务端发送数据：

```java
private boolean isServiceDestroyed=false;

    @Override
    public void onCreate() {
        new Thread(new TcpServer()).start();
        super.onCreate();
    }

    @Override
    public IBinder onBind(Intent intent) {
        throw new UnsupportedOperationException("Not yet implemented");
    }

    private class TcpServer implements Runnable {
        @Override
        public void run() {
            ServerSocket serverSocket;
            try {
                serverSocket=new ServerSocket(8888);
                final Socket client=serverSocket.accept();
                new Thread() {
                    @Override
                    public void run() {
                        responseClient(client);
                    }
                }.start();
            } catch (IOException e) {
                e.printStackTrace();
                return;
            }
        }
    }

    private void responseClient(Socket client) {
        // 用于接收客户端消息
        try {
            InputStream is=client.getInputStream();
            OutputStream os=client.getOutputStream();

            os.write("您好，我是服务端".getBytes());
            os.flush();

            byte[] bytes=new byte[1024];
            int len=0;
            while ((len=is.read(bytes))!=-1) {
                Log.i("MSG", new String(bytes, 0, len));
                os.write(("客户端发送了消息，具体内容:" + new String(bytes, 0, len)).getBytes());
                os.flush();
            }
        } catch (Exception e) {
```

```java
            e.printStackTrace();
        }
    }

    @Override
    public void onDestroy() {
        isServiceDestroyed=true;
        super.onDestroy();
    }
```

客户端接收数据：

```java
private Button bt_send;
    private EditText et_receive;
    private TextView tv_message;
    private List<OutputStream> osList=new ArrayList<>();
    private List<InputStream> isList=new ArrayList<>();


    @Override
    protected void onCreate(Bundle savedInstanceState) {
        super.onCreate(savedInstanceState);
        setContentView(R.layout.activity_socket);
        initView();
        Intent service=new Intent(this, SocketServerService.class);
        startService(service);
        new Thread() {
            @Override
            public void run() {
                connectSocketServer();
            }
        }.start();
    }

    private void initView() {
        et_receive=(EditText) findViewById(R.id.et_receive);
        bt_send=(Button) findViewById(R.id.bt_send);
        tv_message=(TextView) this.findViewById(R.id.tv_message);
        bt_send.setOnClickListener(new View.OnClickListener() {
            // 向服务器发送信息
            @Override
            public void onClick(View v) {
                final String msg=et_receive.getText().toString();
                new Thread() {
                    @Override
                    public void run() {
                        if (!TextUtils.isEmpty(msg) && osList.size() > 0) {
                            OutputStream printWriter=osList.get(0);
                            try {
                                printWriter.write(msg.getBytes());
                            } catch (IOException e) {
                                e.printStackTrace();
                            }
```

```java
                            tv_message.setText(tv_message.getText()+"\n"+
"客户端: " + msg);
                            et_receive.setText("");
                        }
                    }
                }.start();
            }
        });
    }

    private static int num=1;

    private void connectSocketServer() {
        Socket socket=null;
        try {
            tv_message.setText("正在准备第" + num + "次连接,请稍后!");
            Thread.sleep(2000);
            socket=new Socket();
            socket.connect(new InetSocketAddress("localhost", 8888), 1000);
            while (socket.isClosed() && (!socket.isConnected())) {
                socket=new Socket("localhost", 8888);
                tv_message.setText("正在尝试获取连接!");
            }
            tv_message.setText("连接获取成功!");
            InputStream is=socket.getInputStream();
            isList.add(is);

            OutputStream os=socket.getOutputStream();
            osList.add(os);
            byte[] bytes=new byte[1024];
            int len=0;
            while ((len=is.read(bytes))!=-1) {
                Log.i("MSG", new String(bytes, 0, len));
                tv_message.setText(tv_message.getText() + "\n" + "服务端: " +
new String(bytes, 0, len));
            }
        } catch (ConnectException e) {
            if (num++!=5) {
                connectSocketServer();
            }
        } catch (IOException e) {
            e.printStackTrace();
        } catch (InterruptedException e) {
            e.printStackTrace();
        }
    }
```

任务实现

这里将使用安卓移动设备作为服务端,接收可燃气体传感器数据值,当值到达一定程度后,将警告信息发送到客户端的手机上,具体操作步骤如下。

① 登录唯众物联网融合平台，创建项目，在项目下添加设备，添加成功后单击【生成 API】按钮。

② 在 Android Studio 中创建一个新项目，将应用名称设置为 demo_gas_sendMsg，并为项目添加一个空活动。

③ 将项目相关图片复制到 res/drawable 文件夹下，如图 3-15 所示。

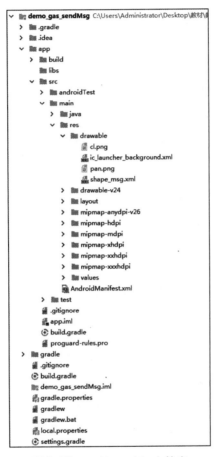

图 3-15　res/drawable 文件夹

④ 切换到【Android】视图，修改 AndroidManifest.xml，添加权限，代码如下。

```
<uses-permission android:name="android.permission.INTERNET"/>
```

⑤ 在资源文件 res/values/colors.xml 中加入字体颜色，代码如下。

```
<resources>
    <color name="colorPrimary">#008577</color>
    <color name="colorPrimaryDark">#00574B</color>
    <color name="colorAccent">#D81B60</color>
</resources>
```

⑥ 添加 dialog_set.xml，在活动布局中添加控件，实现设置阈值弹出框，代码如下。

```xml
<EditText
    android:id="@+id/wnum"
    android:layout_width="200dp"
    android:layout_height="wrap_content"
    android:hint=" 请输入警戒值 "
    android:inputType="number"
    android:paddingLeft="10dp"
    android:paddingRight="10dp"
    android:textColor="#000000" />

<Button
    android:id="@+id/done"
    android:layout_width="wrap_content"
    android:layout_height="wrap_content"
    android:text=" 设置 " />
```

⑦ 修改 activity_main.xml，在主活动布局中添加控件，代码如下，实现效果如图 3-16 所示。

```xml
<android.support.v7.widget.Toolbar
    android:id="@+id/toolbar"
    android:layout_width="match_parent"
    android:layout_height="61dp"
    android:background="?attr/colorPrimary"
    android:minHeight="?attr/actionBarSize"
    android:paddingLeft="20dp"
    android:paddingRight="20dp"
    android:theme="?attr/actionBarTheme">

    <TextView
        android:layout_width="wrap_content"
        android:layout_height="wrap_content"
        android:layout_gravity="center"
        android:text=" 可燃气 "
        android:textSize="24sp"
        android:textStyle="bold"
        tools:layout_editor_absoluteX="264dp"
        tools:layout_editor_absoluteY="14dp" />

    <ImageButton
        android:layout_width="30dp"
        android:layout_height="30dp"
        android:layout_gravity="right"
        android:background="@drawable/cl"
        android:onClick="toSet"
        tools:layout_editor_absoluteX="551dp"
        tools:layout_editor_absoluteY="21dp" />
</android.support.v7.widget.Toolbar>

<LinearLayout
    android:layout_width="match_parent"
    android:layout_height="match_parent"
    android:background="#01123D"
    android:gravity="center"
```

```xml
    android:orientation="vertical">

    <LinearLayout
        android:layout_width="match_parent"
        android:layout_height="0dp"
        android:layout_weight="2"
        android:gravity="center"
        android:orientation="vertical">

        <LinearLayout
            android:layout_width="wrap_content"
            android:layout_height="wrap_content"
            android:background="@drawable/shape_msg"
            android:gravity="center"
            android:orientation="horizontal"
            android:paddingLeft="30dp"
            android:paddingTop="5dp"
            android:paddingRight="30dp"
            android:paddingBottom="5dp">

            <TextView
                android:layout_width="wrap_content"
                android:layout_height="wrap_content"
                android:layout_weight="1"
                android:text="可燃气浓度"
                android:textColor="@android:color/white"
                android:textSize="30sp" />

        </LinearLayout>
    </LinearLayout>

    <LinearLayout
        android:layout_width="match_parent"
        android:layout_height="0dp"
        android:layout_weight="3"
        android:gravity="center"
        android:orientation="vertical">

        <ImageView
            android:id="@+id/imageView"
            android:layout_width="450dp"
            android:layout_height="wrap_content"
            app:srcCompat="@drawable/pan" />

        <TextView
            android:id="@+id/svalue"
            android:layout_width="wrap_content"
            android:layout_height="wrap_content"
            android:layout_marginTop="-240dp"
            android:text="0%"
            android:textColor="#578DFF"
            android:textSize="100sp" />
    </LinearLayout>
</LinearLayout>
```

图 3-16 效果图

⑧ 创建 MyAlertDialog.java 类。自定义弹出框样式，并实现值的设置，代码如下。

```java
public class MyAlertDialog extends AlertDialog {

    private Context context;

    private TextView wnum;
    private Button done;

    private StringBuilder flag;

    protected MyAlertDialog(Context context, StringBuilder flag) {
        super(context);
        this.context=context;
        this.flag=flag;
    }

    @Override
    protected void onCreate(Bundle savedInstanceState) {
        super.onCreate(savedInstanceState);
        // 提前设置Dialog的一些样式
        Window dialogWindow=getWindow();
        dialogWindow.setGravity(Gravity.CENTER);// 设置dialog显示居中
        getWindow().setBackgroundDrawableResource(R.drawable.shape_dialog);
        setContentView(R.layout.dialog_set);
        // 显示键盘
        getWindow().clearFlags(WindowManager.LayoutParams.FLAG_ALT_FOCUSABLE_IM);
        setWidth();
        initView();
    }
```

```java
    private void setWidth() {
        WindowManager windowManager=((Activity) context).getWindowManager();
        Display display=windowManager.getDefaultDisplay();
        WindowManager.LayoutParams lp=getWindow().getAttributes();
        lp.width=display.getWidth()*3/5;// 设置dialog宽度为屏幕的2/3
        getWindow().setAttributes(lp);
    }

    private void initView() {
        wnum=findViewById(R.id.wnum);
        done=findViewById(R.id.done);
        done.setOnClickListener(new View.OnClickListener() {
            @Override
            public void onClick(View v) {
                flag.replace(0, flag.length(), wnum.getText().toString());
                MyAlertDialog.this.dismiss();
                Toast toast=Toast.makeText(context, "设置成功", Toast.LENGTH_SHORT);
                toast.setGravity(Gravity.CENTER, 0, 0);
                toast.show();
            }
        });
    }
}
```

⑨ 修改 MainActivity.java。定义并初始化相关内容,代码如下。

```java
private Wz_HttpTools wht;

    private TextView svalue;
    private MyAlertDialog myAlertDialog;

    private Long startTime;
    private double cct=30;
    private StringBuilder flag=new StringBuilder("");

    private final Timer timer=new Timer();
    svalue=findViewById(R.id.svalue);
    myAlertDialog=new MyAlertDialog(MainActivity.this, flag);
```

⑩ 修改 MainActivity.java。在 onCreate() 方法中的设置 HTTP 请求地址,并启动定时器,每隔 2 s 对 API 接口进行一次请求,代码如下。

```java
wht=new Wz_HttpTools(handler);
        wht.setHttpURL("http://192.168.0.22:8080/wziot/wzIotApi/getOneSensorData/d4e9f9af-1482-4be9-a096-63f7ae015a03/30010");
        TimerTask task=wht.getJsonData();
        timer.schedule(task, 2000, 2000);
```

⑪ 与平台连接成功后,通过 Wz_HttpTools.java 中的 getJsonData() 方法获取数据,代码如下。

```java
    public TimerTask getJsonData() {
        return task=new TimerTask() {
            @Override
            public void run() {
                // TODO Auto-generated method stub
                new Thread(new Runnable() {
                    @Override
                    public void run() {
                        String s=getConnn();
                        if (s!=null) {
                            Message message=new Message();
                            message.what=1;
                            message.obj=s;
                            handler.sendMessage(message);
                        }
                    }
                }).start();
            }
        };
    }
    private String getConnn() {
        HttpURLConnection connection=null;
        BufferedReader reader=null;
        //第一步建立httpurlconnection实例
        try {
            URL url=new URL(httpURL);
            connection=(HttpURLConnection) url.openConnection();
            //自由定制
            connection.setRequestMethod("GET");
            connection.setConnectTimeout(80000);
            connection.setReadTimeout(8000);

            //使用getInputStream获取服务器返回的输入流
            InputStream stresm=connection.getInputStream();
            //对返回的输入流进行读取
            reader=new BufferedReader(new InputStreamReader(stresm));
            StringBuilder response=new StringBuilder();
            String line;
            while ((line=reader.readLine()) != null) {
                response.append(line);
            }

            return response.toString();
        } catch (IOException e) {
            e.printStackTrace();
        } finally {
            if (reader!=null) {
                try {
                    reader.close();
                } catch (IOException e) {
                    e.printStackTrace();
                }
            }
            if (connection!=null) {
```

```
                connection.disconnect();
            }
        }
        return null;
    }
```

⑫ 修改 MainActivity.java。在 sendSMS(String phoneNumber,String message) 方法中，通过短信方式发送信息到指定客户端手机，代码如下。

```
    private void sendSMS(String phoneNumber,String message){
        if(PhoneNumberUtils.isGlobalPhoneNumber(phoneNumber)){
            SmsManager smsm=SmsManager.getDefault();
            ArrayList<String> divideContents=smsm.divideMessage(message);
            ArrayList<PendingIntent> pendingIntents=new ArrayList<>();
            pendingIntents.add(null);
            smsm.sendMultipartTextMessage(phoneNumber, null, divideContents, pendingIntents, null);
        }
    }
```

⑬ 修改 MainActivity.java。在 show(String jsonResult, String index) 方法中，获取可燃气体值，判断后如果超标，将信息发送到客户手机，代码如下。

```
    public void show(String jsonResult, String index) {
        try {
            String code=new JSONObject(jsonResult).getString("code");
            if (!"201".equals(code))
                return;
            JSONArray obj=new JSONObject(jsonResult).getJSONArray("res");
            for (int i=0; i<obj.length(); i++) {
                JSONObject json=(JSONObject) obj.get(i);
                String rindex=json.getString("passGatewayNum");
                if (index.equals(rindex)) {
                    String value=json.getString("value");
                    svalue.setText(value + "%");
                    if (!"".equals(flag.toString()))
                        cct=Integer.parseInt(flag.toString());
                    if (Double.parseDouble(value) > cct) {
                        if (startTime==null) {
                            startTime=System.currentTimeMillis();;
                        } else {
                            if (System.currentTimeMillis()-startTime>30*1000) {
                                startTime=System.currentTimeMillis();
                                sendSMS("手机号码","可燃气体超标了！");
                            }
                        }
                    }
                }
            }
        } catch (JSONException e) {
            e.printStackTrace();
```

物联网移动应用开发

```
        }
    }
```

任务小结

请同学们根据完成情况对完成本次任务的知识、技能等要点进行小结。

任务 3 小结	
知识点掌握情况	
技能点掌握情况	

任务拓展　设置阈值 2

 任务描述

本次任务是通过可燃气体浓度数值的采集，实时检测可燃气体浓度，并和标准值进行比较。用户可根据实际情况自己设置阈值，当高过一定浓度值时进行相应的操作，将警告信息发送到客户端的手机上，并获取短信发送状态。

 任务分析

通过可燃气体浓度数值的采集，可实现当浓度超过阈值时，将警告信息发送到客户端的手机上，并获取短信发送状态。

任务实现

具体操作步骤如下。

① 登录唯众物联网融合平台，创建项目，在项目下添加设备，添加成功后单击【生成 API】按钮。

② 在 Android Studio 中创建一个新项目，将应用名称设置为 demo_gas_sendMsg_ex，并为活动添加一个空活动。

③ 修改 activity_main.xml，在主活动布局中添加控件，实现效果如图 3-17 所示。

```xml
<android.support.v7.widget.Toolbar
    android:id="@+id/toolbar"
    android:layout_width="match_parent"
    android:layout_height="wrap_content"
    android:background="?attr/colorPrimary"
    android:minHeight="?attr/actionBarSize"
    android:paddingLeft="20dp"
```

```xml
        android:paddingRight="20dp"
        android:theme="?attr/actionBarTheme">

        <TextView
            android:layout_width="wrap_content"
            android:layout_height="wrap_content"
            android:layout_gravity="center"
            android:text=" 可燃气 "
            android:textSize="24sp"
            android:textStyle="bold" />

        <ImageButton
            android:layout_width="30dp"
            android:layout_height="30dp"
            android:layout_gravity="right"
            android:background="@drawable/cl"
            android:onClick="toSet"
            tools:layout_editor_absoluteX="551dp"
            tools:layout_editor_absoluteY="23dp" />
</android.support.v7.widget.Toolbar>

<LinearLayout
    android:layout_width="match_parent"
    android:layout_height="match_parent"
    android:background="#01123D"
    android:gravity="center"
    android:orientation="vertical">

    <LinearLayout
        android:layout_width="match_parent"
        android:layout_height="0dp"
        android:layout_weight="2"
        android:gravity="center"
        android:orientation="vertical">

        <LinearLayout
            android:layout_width="wrap_content"
            android:layout_height="wrap_content"
            android:background="@drawable/shape_msg"
            android:gravity="center"
            android:orientation="horizontal"
            android:paddingLeft="30dp"
            android:paddingTop="5dp"
            android:paddingRight="30dp"
            android:paddingBottom="5dp">

            <TextView
                android:layout_width="wrap_content"
                android:layout_height="wrap_content"
                android:layout_weight="1"
                android:text=" 可燃气浓度 "
                android:textColor="@android:color/white"
                android:textSize="30sp" />
```

```xml
        </LinearLayout>
    </LinearLayout>

    <LinearLayout
        android:layout_width="match_parent"
        android:layout_height="0dp"
        android:layout_weight="3"
        android:gravity="center"
        android:orientation="vertical">

        <ImageView
            android:id="@+id/imageView"
            android:layout_width="450dp"
            android:layout_height="wrap_content"
            app:srcCompat="@drawable/pan" />

        <TextView
            android:id="@+id/svalue"
            android:layout_width="wrap_content"
            android:layout_height="wrap_content"
            android:layout_marginTop="-240dp"
            android:text="0%"
            android:textColor="#578DFF"
            android:textSize="100sp" />
    </LinearLayout>

</LinearLayout>
```

图 3-17　效果图

④ 修改 MainActivity.java 类，获取发送短信状态，代码如下。

```java
private Wz_HttpTools wht;

private TextView svalue;
```

```java
    private MyAlertDialog myAlertDialog;

    private Long startTime;
    private double cct=30;
    private StringBuilder flag=new StringBuilder("");

    private PendingIntent sentPI;
    private PendingIntent deliverPI;

    private Toast toast;

    private final Timer timer=new Timer();
    Handler handler=new Handler() {
        @Override
        public void handleMessage(Message msg) {
            String jsonResult=msg.obj.toString();
            // TODO Auto-generated method stub
            // 要做的事情
            switch (msg.what) {
                case 1:
                    show(jsonResult, "1");
                    break;
            }
            super.handleMessage(msg);
        }
    };

    @Override
    protected void onCreate(Bundle savedInstanceState) {
        super.onCreate(savedInstanceState);
        setContentView(R.layout.activity_main);
        initView();
        registerMsgListener();
        wht=new Wz_HttpTools(handler);
        wht.setHttpURL("http://192.168.0.22:8080/wziot/wzIotApi/getOneSensorData/d4e9f9af-1482-4be9-a096-63f7ae015a03/30010");
        TimerTask task=wht.getJsonData();
        timer.schedule(task, 2000, 2000);
    }

    private void registerMsgListener() {
        // 处理返回的发送状态
        String SENT_SMS_ACTION="SENT_SMS_ACTION";
        Intent sentIntent=new Intent(SENT_SMS_ACTION);
        sentPI=PendingIntent.getBroadcast(this, 0, sentIntent,
                0);
        // register the Broadcast Receivers
        this.registerReceiver(new BroadcastReceiver() {
            @Override
            public void onReceive(Context _context, Intent _intent) {
                switch (getResultCode()) {
                    case Activity.RESULT_OK:
                        toast.setText(" 短信发送成功 ");
                        toast.show();
```

```
                        break;
                case SmsManager.RESULT_ERROR_GENERIC_FAILURE:
                        break;
                case SmsManager.RESULT_ERROR_RADIO_OFF:
                        break;
                case SmsManager.RESULT_ERROR_NULL_PDU:
                        break;
            }
        }
    }, new IntentFilter(SENT_SMS_ACTION));

    //处理返回的接收状态
    String DELIVERED_SMS_ACTION="DELIVERED_SMS_ACTION";
    // create the deilverIntent parameter
    Intent deliverIntent=new Intent(DELIVERED_SMS_ACTION);
    deliverPI=PendingIntent.getBroadcast(this, 0,
            deliverIntent, 0);
    this.registerReceiver(new BroadcastReceiver() {
        @Override
        public void onReceive(Context _context, Intent _intent) {
            toast.setText("收信人已经成功接收");
            toast.show();
        }
    }, new IntentFilter(DELIVERED_SMS_ACTION));
}

private void initView() {
    svalue=findViewById(R.id.svalue);
    myAlertDialog=new MyAlertDialog(MainActivity.this, flag);
    toast=Toast.makeText(MainActivity.this, "", Toast.LENGTH_SHORT);
    toast.setGravity(Gravity.CENTER, 0, 0);
}

public void show(String jsonResult, String index) {
    try {
        String code=new JSONObject(jsonResult).getString("code");
        if (!"201".equals(code))
            return;
        JSONArray obj=new JSONObject(jsonResult).getJSONArray("res");
        for (int i=0; i<obj.length(); i++) {
            JSONObject json=(JSONObject) obj.get(i);
            String rindex=json.getString("passGatewayNum");
            if (index.equals(rindex)) {
                String value=json.getString("value");
                svalue.setText(value+"%");
                if (Double.parseDouble(value)>cct) {
                    if (startTime==null) {
                        startTime=System.currentTimeMillis();;
                    } else {
                        if (System.currentTimeMillis()-startTime>30*1000) {
                            startTime=System.currentTimeMillis();
                            sendSMS("15171167443","可燃气体超标了！快跑");
```

```java
                            }
                        }
                    }
                }
            }
        } catch (JSONException e) {
            e.printStackTrace();
        }

    }

    private void sendSMS(String phoneNumber,String message){
        if(PhoneNumberUtils.isGlobalPhoneNumber(phoneNumber)){
            SmsManager smsm=SmsManager.getDefault();
            ArrayList<String> divideContents=smsm.divideMessage(message);
            ArrayList<PendingIntent> pendingIntents=new ArrayList<>();
            ArrayList<PendingIntent> deliveredPendingIntents=new ArrayList<PendingIntent>();
            pendingIntents.add(sentPI);
            deliveredPendingIntents.add(deliverPI);
            smsm.sendMultipartTextMessage(phoneNumber, null, divideContents, pendingIntents, deliveredPendingIntents);
        }
    }

    public void toSet(View view) {
        myAlertDialog.show();
    }
```

项目四

智能家居

项目概述

智能家居又称智慧家居/智能住宅,在国外常用 Smart Home 表示。智能家居是以住宅为平台,兼备建筑、网络通信、信息家电、设备自动化,集系统、结构、服务、管理为一体的高效、舒适、安全、便利、环保的居住环境。

主要包括:

通过物联网技术将家中的各种设备(如音视频设备、照明系统、窗帘控制、空调控制、安防系统、数字影院系统、网络家电以及三表抄送等)连接到一起,提供家电控制、照明控制、窗帘控制、电话远程控制、室内外遥控、防盗报警,以及可编程定时控制等多种功能和手段。

工作任务

- 任务1:智能门锁。
- 任务2:室内光线采集。

学习目标

- 培养学生自主探究和解决问题的能力。
- 培养学生严谨的逻辑思维能力。

项目四 智能家居

任务 1　智能门锁

任务描述

按一下开门按钮就可以打开电锁出门。我们通常称为"单向门禁"或者"单向读卡",这种方式的特点在于便捷,出门方便但安全性略低(防外人不防内部人员),成本更低。大多数办公门禁的内部门都采用这种模式。本次任务是利用安卓平台点击按钮来控制开关门。

通过 Android 应用程序控制门锁的开启和关闭。

知识引入

通过键值控制执行器:

请求方式:GET。

请求地址:http://192.168.0.193:8080/wziot/wzIotApi/controlSensorByKey/{projectId}?uuid={uuid}&key={key}。

URL 请求参数和响应参数如图 4-1 所示。

URL请求参数

参数	类型	描述
projectId	String	项目ID
uuid	String	传感器UUID
key	int	键值

响应参数

参数	类型	描述
code	int	返回状态码
msg	String	返回的消息

图 4-1　URL 请求参数和响应参数

请求示例:

http://192.168.0.193:8080/wziot/wzIotApi/controlSensorByKey/d4e9f9af-1482-4be9-a096-63f7ae015a03?uuid=30015&key=1。

响应示例:

```
{
    "code": 202,
    "msg": "发送控制命令成功",
}
```

通过变量控制执行器:

请求方式：GET。

请求地址：http://192.168.0.193:8080/wziot/wzIotApi/controlSensorByVariable/{projectId}?uuid={uuid}&index={index}&variable={variable}。

URL 请求参数及响应参数如图 4-2 所示。

URL 请求参数		
参数	类型	描述
projectId	String	项目ID
uuid	String	传感器uuid
index	String	索引值
variable	int	变量值
响应参数		
参数	类型	描述
code	int	返回状态码
msg	String	返回的消息

图 4-2　URL 请求参数及响应参数

请求示例：

http://192.168.0.193:8080/wziot/wzIotApi/controlSensorByVariable/d4e9f9af-1482-4be9-a096-63f7ae015a03?uuid=30015&index=1&variable=2。

响应示例

```
{
    "code": 202,
    "msg": "发送控制命令成功",
}
```

任务实现

这里将使用安卓移动设备界面上的按钮和复选框来实现"开启"或"关闭"门禁的功能，具体操作步骤如下。

① 登录唯众物联网融合平台，创建项目，在项目下添加设备，添加成功后单击【生成 API】按钮。

② 在 Android Studio 中创建一个新项目，将应用名称设置为 demo_lock，并为活动添加一个空活动。

③ 将项目相关的图片复制到 res/drawable 文件夹下，如图 4-3 所示。

④ 修改 activity_main.xml，在主活动布局中添加控件，代码如下，实现效果图如图 4-4 所示。

```
<LinearLayout
    android:layout_width="match_parent"
    android:layout_height="0dp"
    android:layout_weight="2"
    android:gravity="center"
```

```xml
        android:orientation="vertical">

        <Button
            android:id="@+id/button"
            android:layout_width="wrap_content"
            android:layout_height="wrap_content"
            android:background="@drawable/shape_msg"
            android:onClick="done"
            android:paddingLeft="30dp"
            android:paddingTop="5dp"
            android:paddingRight="30dp"
            android:paddingBottom="5dp"
            android:text=" 开门 "
            android:textColor="@android:color/white"
            android:textSize="30sp" />
    </LinearLayout>

    <LinearLayout
        android:layout_width="match_parent"
        android:layout_height="0dp"
        android:layout_weight="3"
        android:gravity="center"
        android:orientation="vertical">

        <ImageView
            android:id="@+id/lock"
            android:layout_width="340dp"
            android:layout_height="wrap_content"
            app:srcCompat="@drawable/lock" />
    </LinearLayout>
```

图 4-3　res/drawable 文件夹　　　　　图 4-4　效果图

⑤ 切换到【Android】视图，修改 AndroidManifest.xml，添加权限，代码如下。

```
<uses-permission android:name="android.permission.INTERNET"/>
```

⑥ 修改 MainActivity.java。我们在 MainActivity.java 中增加成员变量，代码如下。

```
private Wz_HttpTools wht;

    private ImageView lock;
    private Timer timer=new Timer();
    private TimerTask task;
```

修改 MainActivity.java。在 onCreate() 方法中的调用的 initView() 方法中初始化内容，代码如下。

```
    private void initView() {
        lock=findViewById(R.id.lock);
    }
```

⑦ 修改 MainActivity.java。在 onCreate() 方法中，设置请求连接，代码如下。

```
wht=new Wz_HttpTools(handler);
wht.setHttpURL("http://192.168.0.230:8080/wziot/wzIotApi/controlSensorByVariable/adeef6b9-c08d-46af-9af6-3d46685ad184?uuid=30014&index=1&variable=1");
```

⑧ 修改 MainActivity.java。添加【开门】按钮方法,将请求发送到物联网融合平台,代码如下。

```
public void done(View view) {
        wht.sendControl();
    }
```

⑨ 当用户将【开启】或【关闭】命令发送到物联网融合平台后，平台返回用户发送状态码以及发送信息，用户根据发送状态码判断加载不同的图片，代码如下。

```
private void show(String jsonResult) {
        try {
            String code=new JSONObject(jsonResult).getString("code");
            if (!"202".equals(code))
                return;
            runOnUiThread(new Runnable() {
                @Override
                public void run() {
                    lock.setBackgroundResource(R.drawable.unlock);
                    if (task!=null)
                        task.cancel();
                    task=new TimerTask() {
                        @Override
                        public void run() {
                            runOnUiThread(new Runnable() {
                                @Override
                                public void run() {
                                    lock.setBackgroundResource(R.drawable.lock);
```

```
                    }
                });
            }
        };
        timer.schedule(task,10000);
    }
});
} catch (JSONException e) {
    e.printStackTrace();
}
```

 任务小结

请同学们根据完成情况对完成本次任务的知识、技能等要点进行小结。

任务 1 小结	
知识点掌握情况	
技能点掌握情况	

任务拓展　RFID 与门锁联动

 任务描述

目前常见的门禁进出方式主要分为两种：一种是刷卡进入，出门时也刷卡。通常称为"双向门禁"或者"双向读卡"，这种方式的优点是安全性级别较高，一般也不准人随便出，出去的人都有记录可以查询，知道谁何时出了哪道门。

 任务分析

通过安卓移动设备来启动门禁系统。当 RFID 启动后，原理与控制照明设备相似，监测门禁卡信息是否正确，正确则开门。

任务实现

具体操作步骤如下。

① 登录唯众物联网融合平台，创建项目，在项目下添加设备，添加成功后单击【生成 API】按钮。

② 在 Android Studio 中创建一个新项目，将应用名称设置为 demo_lock_ex，并为活动添加一个空活动。

③ 将项目相关的图片复制到 res/drawable 文件夹下，如图 4-5 所示。

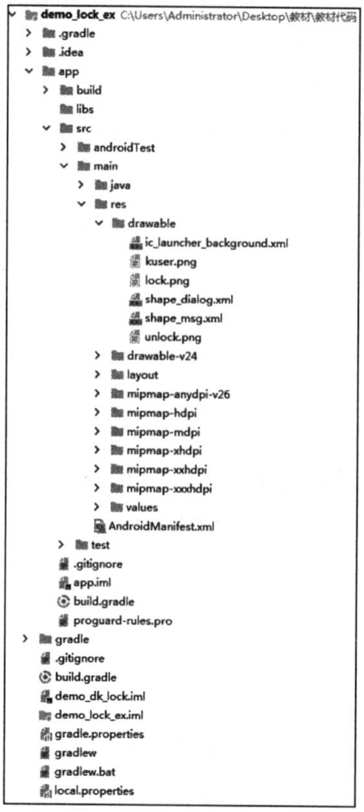

图 4-5 res/drawable 文件夹

④ 创建 dialog_user.xml，代码如下。

```xml
<ImageView
    android:id="@+id/imageView"
    android:layout_width="80dp"
    android:layout_height="wrap_content"
    app:srcCompat="@drawable/kuser" />

<LinearLayout
    android:layout_width="wrap_content"
    android:layout_height="match_parent"
    android:layout_marginLeft="20dp"
    android:orientation="vertical">

    <TextView
        android:id="@+id/uid"
        android:layout_width="wrap_content"
        android:layout_height="0dp"
        android:layout_weight="1"
        android:gravity="center_vertical"
        android:text="用户:"
        android:textColor="#578DFF"
        android:textSize="24sp" />
```

```xml
<TextView
    android:id="@+id/textView2"
    android:layout_width="wrap_content"
    android:layout_height="0dp"
    android:layout_weight="1"
    android:gravity="center"
    android:text=" 打卡成功！ "
    android:textColor="#578DFF"
    android:textSize="24sp" />
</LinearLayout>
```

⑤ 修改 activity_main.xml，在主活动布局中添加控件，代码如下。

```xml
<LinearLayout xmlns:android="http://schemas.android.com/apk/res/android"
    xmlns:app="http://schemas.android.com/apk/res-auto"
    xmlns:tools="http://schemas.android.com/tools"
    android:layout_width="match_parent"
    android:layout_height="match_parent"
    android:background="#01123D"
    android:gravity="center"
    android:orientation="vertical"
    tools:context=".MainActivity">

    <ImageView
        android:id="@+id/lock"
        android:layout_width="340dp"
        android:layout_height="wrap_content"
        app:srcCompat="@drawable/lock" />

</LinearLayout>
```

实现效果图如图 4-6 所示。

图 4-6　效果图

⑥ 修改 MainActivity.java，加入代码如下。

```java
    private Wz_HttpTools wht;
    private Wz_HttpTools lwht;

    private MyAlertDialog myAlertDialog;
    private ImageView lock;

    private final Timer timer=new Timer();
    private TimerTask task;

    private String time;

    Handler handler=new Handler() {
        @Override
        public void handleMessage(Message msg) {
            String jsonResult=msg.obj.toString();
            // TODO Auto-generated method stub
            // 要做的事情
            switch (msg.what) {
                case 1:
                    System.out.println(jsonResult);
                    showRfid(jsonResult);
                    break;
                case 2:
                    showLock(jsonResult);
                    break;
            }
            super.handleMessage(msg);
        }
    };

    @Override
    protected void onCreate(Bundle savedInstanceState) {
        super.onCreate(savedInstanceState);
        setContentView(R.layout.activity_main);
        initView();
        wht=new Wz_HttpTools(handler);
        wht.setHttpURL("http://192.168.0.193:8080/wziot/wzIotApi/getOneSensorData/3179a728-51c4-4fcc-9454-d7324c72187d/30013");
        TimerTask task=wht.getJsonData();
        timer.schedule(task, 2000, 2000);
        lwht=new Wz_HttpTools(handler);
        lwht.setHttpURL("http://192.168.0.230:8080/wziot/wzIotApi/controlSensorByVariable/adeef6b9-c08d-46af-9af6-3d46685ad184?uuid=30014&index=1&variable=1");
    }

    private void initView() {
        myAlertDialog=new MyAlertDialog(MainActivity.this);
        lock=findViewById(R.id.lock);
    }
```

```java
private void showRfid(String jsonResult) {
    try {
        String code=new JSONObject(jsonResult).getString("code");
        if(!"201".equals(code))
            return;
        JSONArray obj=new JSONObject(jsonResult).getJSONArray("res");
        for (int i=0; i<obj.length(); i++) {
            JSONObject json=(JSONObject) obj.get(i);
            String rindex=json.getString("passGatewayNum");
            if("".equals(rindex)) {
                String dtime=json.getString("time");

                if(dtime.equals(time)) {
                    return;
                }

                if(time==null) {
                    time=dtime;
                }

                final String value=json.getString("value");
                runOnUiThread(new Runnable() {
                    @Override
                    public void run() {
                        myAlertDialog.show();
                        myAlertDialog.setText(value);
                        lwht.sendControl();
                    }
                });
            }
        }
    } catch (JSONException e) {
        e.printStackTrace();
    }
}

private void showLock(String jsonResult) {
    try {
        String code=new JSONObject(jsonResult).getString("code");
        if (!"202".equals(code))
            return;
        runOnUiThread(new Runnable() {
            @Override
            public void run() {
                lock.setBackgroundResource(R.drawable.unlock);
                if (task!=null)
                    task.cancel();
                task=new TimerTask() {
                    @Override
                    public void run() {
                        runOnUiThread(new Runnable() {
                            @Override
                            public void run() {
```

```
                                lock.setBackgroundResource(R.drawable.lock);
                            }
                        });
                    }
                };
                timer.schedule(task,10000);
            }
        });
    } catch (JSONException e) {
        e.printStackTrace();
    }
}
```

任务 2 室内光线采集

 任务描述

智能家居中照明控制用处较大,它能在光线不好的时候或门禁开门后监测光线强度,将普通照明手动开与关转换成智能化管理。管理人员或用户只需用安卓移动设备就能对照明控制开关进行设置,减少了工作量。

 **任务分析**

使用安卓移动设备来采集光照传感器的数值。

 知识引入

查询单个传感器最新数据:
请求方式:GET。
请求地址:http://192.168.0.193:8080/wziot/wzIotApi/getOneSensorData/{projectId}/{uuid}。
URL 请求参数及响应参数如图 4-7 所示。

URL请求参数		
参数	类型	描述
projectId	String	项目ID
uuid	String	传感器UUID
响应参数		
参数	类型	描述
code	int	返回状态码
msg	String	返回的消息
res	Collection of SdkSensorData	

图 4-7 URL 请求参数及响应参数

SdkSensor Data 类说明如图 4-8 所示。

参数	类型	描述
uuid	String	传感器UUID
passGatewayNum	String	传感器通道号
value	String	传感器当前通道的值
time	String	数据上传时间戳

图 4-8　SdkSensor Data 类说明

请求示例：
http://192.168.0.193:8080/wziot/wzIotApi/getOneSensorData/d4e9f9af-1482-4be9-a096-63f7ae015a03/30009。

响应示例：

```
{
    "code": 201,
    "msg": "获取数据成功",
    "res": [{
        "passGatewayNum": "1",
        "time": "1556527275637",
        "uuid": "30009",,
        "value": "27"
    }]
}
```

任务实现

这里将使用安卓移动设备界面上的按钮和复选框来实现"打开"或"关闭"照明灯的功能，具体操作步骤如下。

① 登录唯众物联网融合平台，创建项目，在项目下添加设备，添加成功后单击【生成 API】按钮。

② 在 Android Studio 中创建一个新项目，将应用名称设置为 demo_gz，并为活动添加一个空活动。

③ 将项目相关的图片复制到 res/drawable 文件夹下，如图 4-9 所示。

④ 修改 activity_main.xml，在主活动布局中添加控件，代码如下，实现效果如图 4-10 所示。

```
<LinearLayout
```

```xml
android:id="@+id/linearLayout"
android:layout_width="0dp"
android:layout_height="0dp"
android:background="#01123D"
android:gravity="center"
android:orientation="vertical"
app:layout_constraintBottom_toBottomOf="parent"
app:layout_constraintEnd_toEndOf="parent"
app:layout_constraintStart_toStartOf="parent"
app:layout_constraintTop_toTopOf="parent">

<LinearLayout
    android:layout_width="match_parent"
    android:layout_height="0dp"
    android:layout_weight="2"
    android:gravity="center"
    android:orientation="vertical">

    <LinearLayout
        android:layout_width="wrap_content"
        android:layout_height="wrap_content"
        android:background="@drawable/shape_msg"
        android:gravity="center"
        android:orientation="horizontal"
        android:paddingLeft="25dp"
        android:paddingTop="5dp"
        android:paddingRight="25dp"
        android:paddingBottom="5dp">

        <TextView
            android:layout_width="wrap_content"
            android:layout_height="wrap_content"
            android:text=" 当前光照 "
            android:textColor="@android:color/white"
            android:textSize="30sp" />

    </LinearLayout>
</LinearLayout>

<LinearLayout
    android:layout_width="match_parent"
    android:layout_height="0dp"
    android:layout_weight="3"
    android:gravity="center"
    android:orientation="vertical">

    <ImageView
        android:layout_width="450dp"
        android:layout_height="wrap_content"
        app:srcCompat="@drawable/pan" />

    <TextView
        android:id="@+id/svalue"
        android:layout_width="wrap_content"
```

```
            android:layout_height="wrap_content"
            android:layout_marginTop="-240dp"
            android:text="0"
            android:textColor="#4678E2"
            android:textSize="100sp" />
    </LinearLayout>

</LinearLayout>
```

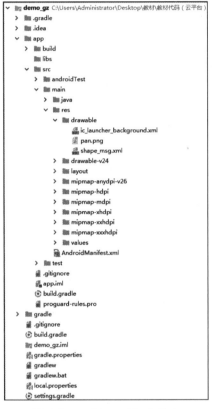

图 4-9 res/drawable 文件夹

图 4-10 效果图

⑤ 切换到【Android】视图，修改 AndroidManifest.xml，添加权限，代码如下。

```
<uses-permission android:name="android.permission.INTERNET"/>
```

⑥ 修改 MainActivity.java。我们在 MainActivity.java 中增加成员变量，代码如下。

```
// 仪表盘数值
private TextView svalue;
// 定时器
private final Timer timer=new Timer();
```

⑦ 修改 MainActivity.java。在 onCreate() 方法中调用的 initView() 方法中初始化内容，代码如下。

```java
private void initView() {
    svalue=findViewById(R.id.svalue);
}
```

⑧ 修改 MainActivity.java。在 onCreate() 方法中调用的 initTools() 设置请求连接，并启动定时器，代码如下。

```java
private void initTools() {
    Wz_HttpTools wht=new Wz_HttpTools(handler);
    wht.setHttpURL("http://192.168.0.22:8080/wziot/wzIotApi/getOneSensorData/adeef6b9-c08d-46af-9af6-3d46685ad184/30009");
    TimerTask task=wht.getJsonData();
    timer.schedule(task, 2000, 2000);
}
```

⑨ 修改 MainActivity.java，在 show(String jsonResult, String index) 方法中接收信息，代码如下。

```java
private void show(String jsonResult, String index) {
    try {
        String code=new JSONObject(jsonResult).getString("code");
        if (!"201".equals(code))
            return;
        // 解析json串，获取信息内容部分
        JSONArray obj=new JSONObject(jsonResult).getJSONArray("res");
        // 循环提取需要的通道号
        for (int i=0; i<obj.length(); i++) {
            JSONObject json=(JSONObject) obj.get(i);
            String rindex=json.getString("passGatewayNum");
            if (index.equals(rindex)) {
                // 获取光敏值
                String value=json.getString("value");
                // 显示光敏到界面
                svalue.setText(value);
            }
        }
    } catch (JSONException e) {
        e.printStackTrace();
    }
}
```

  任务小结

请同学们根据完成情况对完成本次任务的知识、技能等要点进行小结。

任务2小结	
知识点掌握情况	
技能点掌握情况	

任务拓展　光照与灯光联动

**任务描述**

本次任务是通过光照信息数值的采集，实时地将光照值和标准值进行比较，当高过一定值时进行开灯操作，当低于一定值时进行关灯操作。

**任务分析**

采集光照信息，当光照值达到指定值时，开启灯光，当光照低于某个值时，关闭灯光。

任务实现

具体操作步骤如下。

① 登录唯众物联网融合平台，创建项目，在项目下添加设备，添加成功后单击【生成 API】按钮。

② 在 Android Studio 中创建一个新项目，将应用名称设置为 demo_gz_ex，并为活动添加一个空活动。

③ 将项目相关的图片复制到 res/drawable 文件夹下。

```
<LinearLayout xmlns:android="http://schemas.android.com/apk/res/android"
    xmlns:tools="http://schemas.android.com/tools"
    android:layout_width="match_parent"
    android:layout_height="wrap_content"
    android:gravity="center"
    android:orientation="horizontal"
    android:paddingLeft="40dp"
    android:paddingTop="10dp"
    android:paddingRight="40dp"
    android:paddingBottom="10dp"
    android:visibility="visible">

    <EditText
        android:id="@+id/wnum"
        android:layout_width="200dp"
        android:layout_height="wrap_content"
        android:hint="请输入警戒值"
        android:inputType="number"
        android:paddingLeft="10dp"
        android:paddingRight="10dp"
        android:textColor="#000000" />

    <Button
        android:id="@+id/done"
        android:layout_width="wrap_content"
```

```
        android:layout_height="wrap_content"
        android:text=" 设置 " />

</LinearLayout>
```

④ 修改 activity_main.xml，在主活动布局中添加控件，代码如下。

```
<android.support.v7.widget.Toolbar
    android:id="@+id/toolbar"
    android:layout_width="match_parent"
    android:layout_height="wrap_content"
    android:background="?attr/colorPrimary"
    android:minHeight="?attr/actionBarSize"
    android:paddingLeft="20dp"
    android:paddingRight="20dp"
    android:theme="?attr/actionBarTheme">

    <TextView
        android:layout_width="wrap_content"
        android:layout_height="wrap_content"
        android:layout_gravity="center"
        android:text=" 光照强度 "
        android:textColor="@android:color/white"
        android:textSize="24sp"
        android:textStyle="bold" />

    <ImageButton
        android:layout_width="30dp"
        android:layout_height="30dp"
        android:layout_gravity="right"
        android:background="@drawable/cl"
        android:onClick="toSet"
        android:contentDescription="TODO" />
</android.support.v7.widget.Toolbar>

<LinearLayout
    android:id="@+id/linearLayout"
    android:layout_width="match_parent"
    android:layout_height="0dp"
    android:layout_weight="2"
    android:background="#01123D"
    android:gravity="center"
    android:orientation="vertical"
    app:layout_constraintBottom_toBottomOf="parent"
    app:layout_constraintEnd_toEndOf="parent"
    app:layout_constraintStart_toStartOf="parent"
    app:layout_constraintTop_toTopOf="parent">

    <LinearLayout
        android:layout_width="match_parent"
        android:layout_height="0dp"
        android:layout_weight="2"
        android:gravity="center"
```

```xml
    android:orientation="vertical">

    <LinearLayout
        android:layout_width="wrap_content"
        android:layout_height="wrap_content"
        android:background="@drawable/shape_msg"
        android:gravity="center"
        android:orientation="horizontal"
        android:paddingLeft="25dp"
        android:paddingTop="5dp"
        android:paddingRight="25dp"
        android:paddingBottom="5dp">

        <TextView
            android:layout_width="wrap_content"
            android:layout_height="wrap_content"
            android:text=" 当前光照 "
            android:textColor="@android:color/white"
            android:textSize="30sp" />

        <ImageView
            android:id="@+id/deng"
            android:layout_width="35dp"
            android:layout_height="wrap_content"
            android:layout_marginLeft="10dp"
            android:layout_weight="1"
            android:src="@drawable/deng" />

    </LinearLayout>
</LinearLayout>

<LinearLayout
    android:layout_width="match_parent"
    android:layout_height="0dp"
    android:layout_weight="3"
    android:gravity="center"
    android:orientation="vertical">

    <ImageView
        android:layout_width="450dp"
        android:layout_height="wrap_content"
        app:srcCompat="@drawable/pan" />

    <TextView
        android:id="@+id/svalue"
        android:layout_width="wrap_content"
        android:layout_height="wrap_content"
        android:layout_marginTop="-240dp"
        android:text="0"
        android:textColor="#4678E2"
        android:textSize="100sp" />
</LinearLayout>

</LinearLayout>
```

实现效果如图 4-11 所示。

图 4-11　效果图

⑤ 修改 MainActivity.java，加入代码如下。

```
private Wz_HttpTools wht;
private Wz_HttpTools lwht;
private Wz_HttpTools fwht;

// 仪表盘数值
private TextView svalue;
private ImageView deng;
private MyAlertDialog myAlertDialog;

private boolean hasLight=false;
private double cct=30;
private StringBuilder flag=new StringBuilder("");

// 定时器
private final Timer timer=new Timer();
Handler handler=new Handler() {
    @Override
    public void handleMessage(Message msg) {
        String jsonResult=msg.obj.toString();
        // TODO Auto-generated method stub
        // 要做的事情
        switch (msg.what) {
            case 1:
                show(jsonResult,"1");
```

```
                break;
            case 2:
                showDeng(jsonResult);
                break;
        }
    }
};


@Override
protected void onCreate(Bundle savedInstanceState) {
    super.onCreate(savedInstanceState);
    setContentView(R.layout.activity_main);
    initView();
    initTools();
}

private void initTools() {
    wht=new Wz_HttpTools(handler);
    wht.setHttpURL("http://192.168.0.22:8080/wziot/wzIotApi/getOneSensorData/adeef6b9-c08d-46af-9af6-3d46685ad184/30009");
    TimerTask task=wht.getJsonData();
    timer.schedule(task, 2000, 2000);
    lwht=new Wz_HttpTools(handler);
    lwht.setHttpURL("http://192.168.0.230:8080/wziot/wzIotApi/controlSensorByVariable/adeef6b9-c08d-46af-9af6-3d46685ad184?uuid=30015&index=1&variable=1");
    fwht=new Wz_HttpTools(handler);
    fwht.setHttpURL("http://192.168.0.230:8080/wziot/wzIotApi/controlSensorByVariable/adeef6b9-c08d-46af-9af6-3d46685ad184?uuid=30015&index=1&variable=2");
}

private void initView() {
    svalue=findViewById(R.id.svalue);
    deng=findViewById(R.id.deng);
    myAlertDialog=new MyAlertDialog(MainActivity.this, flag);
}

private void show(String jsonResult, String index) {
        try {
            String code=new JSONObject(jsonResult).getString("code");
            if (!"201".equals(code))
                return;
            // 解析json串，获取信息内容部分
            JSONArray obj=new JSONObject(jsonResult).getJSONArray("res");
            // 循环提取需要的通道号
            for (int i=0; i<obj.length(); i++) {
                JSONObject json=(JSONObject) obj.get(i);
                String rindex=json.getString("passGatewayNum");
                if (index.equals(rindex)) {
                    // 获取光敏值
                    String value=json.getString("value");
                    // 显示光敏值到界面
```

```
                        svalue.setText(value);
                        if (!"".equals(flag.toString()))
                            cct=Integer.parseInt(flag.toString());
                        if(Double.parseDouble(value) > cct) {
                            if(!hasLight) {
                                hasLight=true;
                                lwht.sendControl();
                            }
                        } else {
                            if (hasLight) {
                                hasLight=false;
                                fwht.sendControl();
                            }
                        }
                    }
                }
        } catch (JSONException e) {
            e.printStackTrace();
        }
    }

    private void showDeng(String jsonResult) {
        try {
            String code=new JSONObject(jsonResult).getString("code");
            if(!"202".equals(code))
                return;
            if(hasLight) {
                deng.setBackgroundResource(R.drawable.deng_on);
            } else {
                deng.setBackgroundResource(R.drawable.deng);
            }
        } catch (JSONException e) {
            e.printStackTrace();
        }
    }

    public void toSet(View view) {
        myAlertDialog.show();
    }
```

项目五

环境数据存储与查询

项目概述

在生活中很多数据需要保存,以便查询或修改。在这里我们简单地研究温度的保存。
主要包括:SQLite 数据库的使用等。

工作任务

- 任务 1:温度值数据存储。
- 任务 2:温度值历史查询。

学习目标

- 培养学生自主探究和解决问题的能力。
- 培养学生严谨的逻辑思维能力。

任务 1　温度值数据存储

 任务描述

在农业、工业、家庭等环境下我们可能需要记录一定量的温度数据,人工记录数据容易出错,整理数据也不方便。

本次任务即安卓系统中的数据库的使用。

 任务分析

通过 SQLite 存储实时采集的温度值及时间。

 知识引入

1. SQLite 介绍

SQLite 的功能强大，在数据存储、管理、维护等各方面都相当出色。其具有以下特点：

① 轻量级。只需要带一个动态库，就可以使用它的全部功能，而且那个动态库的尺寸相当小。

② 独立性。数据库的核心引擎不需要依赖第三方软件，也不需要所谓的"安装"。

③ 隔离性。数据库中所有的信息（比如表、视图、触发器等）都包含在一个文件夹内，方便管理和维护。

④ 跨平台。目前支持大部分操作系统，不只计算机操作系统，其在众多的手机系统中也可运行，如 Android 等。

⑤ 多语言接口。数据库支持多语言编程接口。

⑥ 安全性。通过独占性和共享锁来实现独立事务处理。这意味着多个进程可以在同一时间从同一数据库读取数据，但只能有一个可以写入数据。

2. SQLite 数据库创建

SQLiteOpenHelper 是 SQLiteDatabase 的一个帮助类，用来管理数据库的创建和版本的更新。

getReadableDatabase()：创建或打开一个只读数据库。

getWritableDatabase()：创建或打开一个读写数据库。

创建 MyDataBaseUtil 类，继承 SQLiteOpenHelper，用于数据库的创建以及修改，代码如下。

```
public class MyDataBaseUtil extends SQLiteOpenHelper {
    public MyDataBaseUtil(Context context, String databaseName, SQLiteDatabase.CursorFactory factory, int version) {
        super(context, databaseName, factory, version);
    }

    @Override
    public void onCreate(SQLiteDatabase db) {
        // 此方法在第一次创建数据库时执行 (是否执行根据  databaseName.db  文件存在与否决定)
    }

    @Override
    public void onUpgrade(SQLiteDatabase db, int oldVersion, int newVersion) {

    }
```

在 MainActivity.java 类中，创建 initDataBase() 方法，初始化数据库信息，代码如下。

```
private SQLiteDatabase db;
private void initDataBase() {
    MyDataBaseUtil dataBaseUtil=new MyDataBaseUtil(MainActivity.this, "user_db", null, 1);
```

}

3. 数据表创建

修改 initDataBase() 方法，加入创建表代码，代码如下。

```
// 每次启动项目就重新创建表，根据自身需求决定
    String drop="DROP TABLE IF EXISTS 'user'";
    db=dataBaseUtil.getReadableDatabase();
    db.execSQL(drop);
    // 创建表，如果不需要每次启动项目重新创建，可将该段代码添加到 MyDataBaseUtil.onCreate 中
    String create="CREATE TABLE user(id VARCHAR(50),username VARCHAR(200));";
    db.execSQL(create);

    //getReadableDatabase() 及 getWritableDatabase 在数据库第一次创建时将会调用 MyDataBaseUtil.onCreate 方法
    db=dataBaseUtil.getWritableDatabase();
    dbMap.put("db", db);
    dbMap.put("handler", handler);
```

4. 数据表操作（增删改查）

public int delete(String table,String whereClause,String[] whereArgs)：删除数据行的便捷方法。

Public long insert(String table,String nullColumnHack,ContentValues values)：添加数据行的便捷方法。

public int update(String table, ContentValues values, String whereClause, String[] whereArgs)：更新数据行的便捷方法。

public void execSQL(String sql)：执行一个 SQL 语句，可以是一个 select 或其他的 SQL 语句。

public void close()：关闭数据库。

public Cursor query(String table, String[] columns, String selection, String[] selectionArgs, String groupBy, String having, String orderBy, String limit)：查询指定的数据表返回一个带游标的数据集。

public Cursor rawQuery(String sql, String[] selectionArgs)：运行一个预置的 SQL 语句，返回带游标的数据集。（与上面的语句最大的区别就是防止 SQL 注入。）

 任务实现

这里将使用安卓移动设备界面上的 4 个按钮实现"增、删、改、查"的功能，具体操作步骤如下。

① 登录唯众物联网融合平台，创建项目，在项目下添加设备，添加成功后单击【生成 API】按钮。

② 在 Android Studio 中创建一个新项目，将应用名称设置为 Android-SQLiteWD，并为活动添加一个空活动。

③ 切换到【Android】视图，修改 AndroidManifest.xml，添加权限，代码如下。

```xml
<uses-permission android:name="android.permission.INTERNET"/>
```

④ 在资源文件 res/values/colors.xml 中加入字体颜色,代码如下。

```xml
<resources>
    <color name="colorPrimary">#008577</color>
    <color name="colorPrimaryDark">#00574B</color>
    <color name="colorAccent">#D81B60</color>
</resources>
```

⑤ 修改 activity_main.xml,在主活动布局中添加控件,代码如下。

```xml
<TextView
    android:layout_width="match_parent"
    android:layout_height="wrap_content"
    android:text="每2s获取的信息"
    android:textAlignment="center" />

<LinearLayout
    android:layout_width="match_parent"
    android:layout_height="wrap_content"
    android:orientation="horizontal">

    <TextView
        android:id="@+id/value_tv"
        android:layout_width="1dp"
        android:layout_height="wrap_content"
        android:layout_weight="0.5"
        android:text="温度值:"
        android:textAlignment="center" />

    <TextView
        android:id="@+id/createTime_tv"
        android:layout_width="1dp"
        android:layout_height="wrap_content"
        android:layout_weight="0.5"
        android:text="获取时间:"
        android:textAlignment="center" />
</LinearLayout>
```

⑥ 根据知识引入内容创建 MyDataBaseUtil.java 类,代码如下。

```java
public class MyDataBaseUtil extends SQLiteOpenHelper {
    public MyDataBaseUtil(Context context, String databaseName, SQLiteDatabase.
CursorFactory factory, int version) {
        super(context, databaseName, factory, version);
    }

    @Override
    public void onCreate(SQLiteDatabase db) {
        // 此方法在第一次创建数据库时执行(是否执行根据 databaseName.db 文件存在与否决定)
    }
```

实现效果如图 5-1 所示。

图 5-1 效果图

```
    @Override
    public void onUpgrade(SQLiteDatabase db, int oldVersion, int newVersion) {
    }
}
```

⑦ 修改 MainActivity.java，实现效果，代码如下。

```
public class MainActivity extends AppCompatActivity {
    private SQLiteDatabase db;
    private Integer id=0;
    private Integer lastId=0;

    private TextView createTime_tv;
    private TextView value_tv;

    @Override
    protected void onCreate(Bundle savedInstanceState) {
        super.onCreate(savedInstanceState);
        setContentView(R.layout.activity_main);
        initView();
        initDataBase();
        initTools();
    }

    private void initView() {
        value_tv=(TextView) findViewById(R.id.value_tv);
        createTime_tv=(TextView) findViewById(R.id.createTime_tv);
    }
```

```java
    private final Timer timer=new Timer();
    Handler handler=new Handler() {
        @Override
        public void handleMessage(Message msg) {
            String jsonResult=msg.obj.toString();
            switch (msg.what) {
                case 1:
                    saveData(jsonResult, "1");//index: 通道号，通常为1
                    break;
            }
            super.handleMessage(msg);
        }
    };

    private void initDataBase() {
        MyDataBaseUtil dataBaseUtil=new MyDataBaseUtil(MainActivity.this,
"wd_db", null, 1);

        // 每次启动项目就重新创建表，根据自身需求决定
        String drop="DROP TABLE IF EXISTS 'temperature'";
        db=dataBaseUtil.getReadableDatabase();
        db.execSQL(drop);

        // 创建表，如果不需要每次启动项目时重新创建，可将该段代码添加到MyDataBaseUtil.
onCreate 中
        String create="CREATE TABLE temperature(id INTEGER PRIMARY KEY
AUTOINCREMENT,value VARCHAR(200),createTime VARCHAR(200));";
        db.execSQL(create);

        //getReadableDatabase() 及 getWritableDatabase 在数据库第一次创建时将会调用
MyDataBaseUtil.onCreate 方法
        db=dataBaseUtil.getWritableDatabase();
    }

    private void initTools() {
        Wz_HttpTools wht=new Wz_HttpTools(handler);
        wht.setHttpURL("http://192.168.0.193:8080/wziot/wzIotApi/
getOneSensorData/3179a728-51c4-4fcc-9454-d7324c72187d/30008");
        TimerTask task=wht.getJsonData();
        timer.schedule(task, 2000, 2000);// 每隔2 s 采集一次数据
    }

    // 将采集到的数据进行存储
    private void saveData(String jsonResult, String index) {
        try {
            // 解析json串，获取信息内容部分
            JSONArray obj=new JSONObject(jsonResult).getJSONArray("res");
            // 循环提取需要的通道号
            for (int i=0; i<obj.length(); i++) {
                JSONObject json=(JSONObject) obj.get(i);
                String rindex=json.getString("passGatewayNum");
                String time=json.getString("time");
                if (index.equals(rindex)) {// 获取温度值
                    String value=json.getString("value");
```

项目五　环境数据存储与查询

```
                    ContentValues cv=new ContentValues();
                    cv.put("value", value);
                    cv.put("createTime", time);
                    long resultRow=db.insert("temperature", null, cv);
                    value_tv.setText("温度值：" + value);
                    createTime_tv.setText("获取时间：" + time);
                }
            }
        } catch (JSONException e) {
            e.printStackTrace();
        }
    }
}
```

 任务小结

请同学们根据完成情况对完成本次任务的知识、技能等要点进行小结。

任务 1 小结	
知识点掌握情况	
技能点掌握情况	

任务拓展　SQLite 数据库删除和更新

 任务描述

本次任务是对 SQLite 数据库中指定数据进行删除，并对数据库中指定数据进行更新，最后将更新结果显示给用户。

任务分析

对上述任务中创建的 SQLite 数据库中数据进行删除和更新操作。

任务实现

具体操作步骤如下。

① 登录唯众物联网融合平台，创建项目，在项目下添加设备，添加成功后单击【生成 API】按钮。

② 在 Android Studio 中创建一个新项目，将应用名称设置为 Android-SQLiteWD-CRUD，并为活动添加一个空活动。

③ 创建 listview_impl.xml.xml，自定义 list 格式，代码如下。

图 5-2　效果图

```
<LinearLayout
    android:id="@+id/listView_ll"
    android:layout_width="match_parent"
    android:layout_height="wrap_content"
    android:orientation="horizontal"
    android:weightSum="1">

    <TextView
        android:id="@+id/id_tv"
        android:layout_width="1dp"
        android:layout_height="wrap_content"
        android:layout_weight="0.15"
        android:textAlignment="center" />

    <EditText
        android:id="@+id/value_et"
        android:layout_width="1dp"
        android:layout_height="wrap_content"
        android:layout_weight="0.15"
        android:textAlignment="center" />

    <TextView
        android:id="@+id/time_tv"
        android:layout_width="1dp"
        android:layout_height="wrap_content"
        android:layout_weight="0.3"
        android:textAlignment="center" />
```

```xml
<Button
    android:id="@+id/update_bt"
    android:layout_width="1dp"
    android:layout_height="wrap_content"
    android:layout_weight="0.2"
    android:onClick="onClick"
    android:text=" 修改 " />

<Button
    android:id="@+id/delete_bt"
    android:layout_width="1dp"
    android:layout_height="wrap_content"
    android:layout_weight="0.2"
    android:onClick="onClick"
    android:text=" 删除 " />
</LinearLayout>
```

④ 修改 activity_main.xml，在主活动布局中添加控件，代码如下，实现效果如图 5-2 所示。

```xml
<TextView
    android:layout_width="match_parent"
    android:layout_height="wrap_content"
    android:text=" 室内环境数据存储 "
    android:textAlignment="center"
    android:textSize="25dp" />

<LinearLayout
    android:layout_width="match_parent"
    android:layout_height="wrap_content"
    android:orientation="horizontal">

    <LinearLayout
        android:layout_width="wrap_content"
        android:layout_height="wrap_content">

        <TextView
            android:layout_width="wrap_content"
            android:layout_height="wrap_content"
            android:text=" 每 2 秒获取的信息 ==>"
            android:textAlignment="textStart" />
    </LinearLayout>

    <LinearLayout
        android:layout_width="match_parent"
        android:layout_height="wrap_content"
        android:orientation="horizontal">

        <TextView
            android:id="@+id/value_tv"
            android:layout_width="1dp"
            android:layout_height="wrap_content"
            android:layout_weight="0.4"
```

```xml
            android:text=" 温度值 :"
            android:textAlignment="textStart" />

        <TextView
            android:id="@+id/createTime_tv"
            android:layout_width="1dp"
            android:layout_height="wrap_content"
            android:layout_weight="0.6"
            android:text=" 采集时间 :"
            android:textAlignment="textStart" />
    </LinearLayout>
</LinearLayout>

<Button
    android:id="@+id/showData_bt"
    android:layout_width="match_parent"
    android:layout_height="wrap_content"
    android:onClick="onClick"
    android:text=" 点击加载采集到的数据 " />

<LinearLayout
    android:id="@+id/title_layoutLL"
    android:layout_width="match_parent"
    android:layout_height="wrap_content"
    android:orientation="horizontal"
    android:weightSum="1">

    <TextView
        android:layout_width="1dp"
        android:layout_height="wrap_content"
        android:layout_weight="0.15"
        android:text=" 数据 id"
        android:textAlignment="center" />

    <TextView
        android:layout_width="1dp"
        android:layout_height="wrap_content"
        android:layout_weight="0.15"
        android:text=" 温度值 "
        android:textAlignment="center" />

    <TextView
        android:layout_width="1dp"
        android:layout_height="wrap_content"
        android:layout_weight="0.3"
        android:text=" 采集时间 "
        android:textAlignment="center" />

    <TextView
        android:layout_width="1dp"
        android:layout_height="wrap_content"
        android:layout_weight="0.4"
        android:text=" 操作 "
        android:textAlignment="center" />
```

```xml
</LinearLayout>

<ListView
    android:id="@+id/temp_lv"
    android:layout_width="match_parent"
    android:layout_height="wrap_content">

</ListView>
```

⑤ 创建 MyAdapter.java 类，代码如下。

```java
public class MyAdapter extends ArrayAdapter<Temperature> {
    public MyAdapter(Context context, int resource, List objects) {
        super(context, resource, objects);
    }

    @Override
    public View getView(int position, View convertView, ViewGroup parent) {
        Temperature temperature=(Temperature) getItem(position);
        View view=LayoutInflater.from(getContext()).inflate(R.layout.listview_impl, null);

        // 获取组件
        TextView id_tv=(TextView) view.findViewById(R.id.id_tv);
        TextView time_tv=(TextView) view.findViewById(R.id.time_tv);
        final EditText value_et=(EditText) view.findViewById(R.id.value_et);
        Button update=view.findViewById(R.id.update_bt);
        Button delete=view.findViewById(R.id.delete_bt);

        // 设置值
        id_tv.setText(temperature.getId().toString());
        value_et.setText(temperature.getValue());
        time_tv.setText(temperature.getCreateTime());

        final Integer tempId=temperature.getId();

        // 设置点击事件
        update.setOnClickListener(new View.OnClickListener() {
            @Override
            public void onClick(View v) {
                SQLiteDatabase db=(SQLiteDatabase) MainActivity.dbMap.get("db");
                Handler handler=(Handler) MainActivity.dbMap.get("handler");
                ContentValues cv=new ContentValues();
                cv.put("value", value_et.getText().toString());
                int resultRow=db.update("temperature", cv, "id=?", new String[]{tempId.toString()});
                if (resultRow>0) {
                    Message message=new Message();
                    message.what=2;
                    message.obj=" 修改 id="+tempId+" 的数据成功！";
                    handler.sendMessage(message);
                }
```

```
                }
            });
            // 设置点击事件
            delete.setOnClickListener(new View.OnClickListener() {
                @Override
                public void onClick(View v) {
                    SQLiteDatabase db=(SQLiteDatabase) MainActivity.dbMap.get("db");
                    Handler handler=(Handler) MainActivity.dbMap.get("handler");
                    int resultRow=db.delete("temperature", "id=?", new String[]{tempId.toString()});
                    if (resultRow>0) {
                        Message message=new Message();
                        message.what=2;
                        message.obj=" 删除 id=" + tempId +" 的数据成功!";
                        handler.sendMessage(message);
                    }
                }
            });
            return view;
        }
    }
```

⑥ 修改 MainActivity.java，实现效果，代码如下。

```
public static HashMap<String, Object> dbMap=new HashMap<>();

private SQLiteDatabase db;

private TextView value_tv;
private TextView createTime_tv;
private Button showData_bt;
private LinearLayout title_layoutLL;

private SimpleDateFormat sdf=new SimpleDateFormat("yyyy/MM/dd HH:mm:ss");

@Override
protected void onCreate(Bundle savedInstanceState) {
    super.onCreate(savedInstanceState);
    setContentView(R.layout.activity_main);

    //1.初始化组件
    initView();

    //2.初始化数据库
    initDataBase();

    //3.初始化请求
    initTools();

}

private void initView() {
    value_tv=(TextView) findViewById(R.id.value_tv);
```

```java
        createTime_tv=(TextView) findViewById(R.id.createTime_tv);
        title_layoutLL=(LinearLayout) findViewById(R.id.title_layoutLL);
        showData_bt=(Button) findViewById(R.id.showData_bt);
    }

    private final Timer timer=new Timer();
    Handler handler=new Handler() {
        @Override
        public void handleMessage(Message msg) {
            String jsonResult=msg.obj.toString();
            switch (msg.what) {
                case 1:
                    saveData(jsonResult, "1");//index:通道号,通常为1
                    break;
                case 2:
                    Toast.makeText(MainActivity.this, msg.obj.toString(), Toast.LENGTH_SHORT).show();
                    initListView();
                    break;
                case 3:
                    Toast.makeText(MainActivity.this, msg.obj.toString(), Toast.LENGTH_SHORT).show();
                    break;
            }
            super.handleMessage(msg);
        }
    };

    private void initDataBase() {
        MyDataBaseUtil dataBaseUtil=new MyDataBaseUtil(MainActivity.this, "wd_db", null, 1);

        // 每次启动项目就重新创建表,根据自身需求决定
        /*String drop="DROP TABLE IF EXISTS 'temperature'";
        db=dataBaseUtil.getReadableDatabase();
        db.execSQL(drop);*/

        // 创建表,如果不需要每次启动项目时重新创建,可将该段代码添加到MyDataBaseUtil.onCreate中
        /*String create="CREATE TABLE temperature(id INTEGER PRIMARY KEY AUTOINCREMENT,value VARCHAR(200),createTime VARCHAR(200));";
        db.execSQL(create);*/

        //getReadableDatabase()及getWritableDatabase在数据库第一次创建时将会调用MyDataBaseUtil.onCreate方法
        db=dataBaseUtil.getWritableDatabase();
        dbMap.put("db", db);
        dbMap.put("handler", handler);
    }

    private void initTools() {
        Wz_HttpTools wht=new Wz_HttpTools(handler);
        wht.setHttpURL("http://192.168.0.193:8080/wziot/wzIotApi/getOneSensorData/3179a728-51c4-4fcc-9454-d7324c72187d/30008");
        TimerTask task=wht.getJsonData();
```

```java
        timer.schedule(task, 2000, 5000);// 每隔2s采集一次数据
}

// 将采集到的数据进行存储
private void saveData(String jsonResult, String index) {
    try {
        // 解析json串，获取信息内容部分
        JSONArray obj=new JSONObject(jsonResult).getJSONArray("res");
        // 循环提取需要的通道号
        for (int i=0; i<obj.length(); i++) {
            JSONObject json=(JSONObject) obj.get(i);
            String rindex=json.getString("passGatewayNum");
            if (index.equals(rindex)) {// 获取温度值
                String time=json.getString("time");
                String value=json.getString("value");
                ContentValues cv=new ContentValues();
                cv.put("value", value);
                cv.put("createTime", time);
                value_tv.setText("温度值:" + value);
                createTime_tv.setText("采集时间:" + time);
                long resultRow=db.insert("temperature", null, cv);
                if (resultRow>0) {
                    Message message=new Message();
                    message.what=3;
                    message.obj="采集到了一条数据 \n 温度值为:" + value + "\n 采
集时间为:" + sdf.format(Long.valueOf(time));
                    handler.sendMessage(message);
                }
            }
        }
    } catch (JSONException e) {
        e.printStackTrace();
    }
}

private void initListView() {
    List<Temperature> temperatureList=new ArrayList<Temperature>();
    Cursor cursor;
    // 获取数据添加到集合中
    cursor=db.query("temperature", new String[]{"id", "value",
"createTime"}, null, null, null, null, "id desc");
    while (cursor.moveToNext()) {
        Integer id=cursor.getInt(cursor.getColumnIndex("id"));
        String value=cursor.getString(cursor.getColumnIndex("value"));
        String createTime=cursor.getString(cursor.
getColumnIndex("createTime"));
        temperatureList.add(new Temperature(id, value, sdf.format(Long.
valueOf(createTime))));
    }
    if (temperatureList.size()<=0) {
        title_layoutLL.setVisibility(View.GONE);
    } else {
        title_layoutLL.setVisibility(View.VISIBLE);
    }
```

```
    // 初始化适配器并进行ListView UI 渲染
    MyAdapter adapter=new MyAdapter(MainActivity.this, R.layout.listview_
impl, temperatureList);
    ListView listView=(ListView) findViewById(R.id.temp_lv);
    listView.setAdapter(adapter);
}

@Override
public void onClick(View v) {
    switch (v.getId()) {
        case R.id.showData_bt: {
            initListView();
        }
    }
}
```

任务 2　温度值历史查询

 任务描述

温度的数据样本存好后可随时根据需要来查询。本次任务需利用 SQLite 数据库，查询并显示需要的数据。

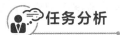 **任务分析**

设置查询时间段，查询 SQLite 存储的温度值，并以列表形式显示在主界面。

 知识引入

1. 时间控件

Android 的自带时间选择控件，是一个让用户既能输入又能选择的控件。其可多点点击选择日期和时间。

① 日期控件 DatePicker。

② 时间控件 TimePicker。

```
<DatePicker
    android:id="@+id/date_Picker"
    android:layout_width="wrap_content"
    android:layout_height="350dp"
    android:layout_gravity="center_horizontal"/>

<TimePicker
    android:id="@+id/time_Picker"
    android:layout_width="wrap_content"
    android:layout_height="350dp"
```

```
android:layout_gravity="center_horizontal"
android:layout_marginLeft="30dp"/>
```

2. 列表控件

ListView 是 Android 中最重要的组件之一，几乎每个 Android 应用中都会使用 ListView。它以垂直列表的方式列出所需的列表项。

```
java.lang.Object
    android.view.View
     android.view.ViewGroup
       android.widget.AdapterView
         android.widget.AbsListView
           android.widget.ListView
```

（1）ListView 的两个职责

① 将数据填充到布局。

② 处理用户的选择点击等操作。

（2）列表的显示需要三个元素

① ListVeiw：用来展示列表的 View。

② 适配器：用来把数据映射到 ListView 上的中介。

③ 数据源：具体的将被映射的字符串、图片，或者基本组件。

（3）什么是适配器

适配器是一个连接数据和 AdapterView 的桥梁，通过它能有效地实现数据与 AdapterView 的分离设置，使 AdapterView 与数据的绑定更加简便，修改更加方便。将数据源的数据适配到 ListView 中的常用适配器有：ArrayAdapter、SimpleAdapter 和 SimpleCursorAdapter。

① ArrayAdapter 最为简单，只能展示一行字。

② SimpleAdapter 有最好的扩充性，可以自定义各种各样的布局，除了文本外，还可以放 ImageView（图片）、Button（按钮）、CheckBox（复选框）等等。

③ SimpleCursorAdapter 可以认为是 SimpleAdapter 对数据库的简单结合，可以方便地把数据库的内容以列表的形式展示出来。

但是实际工作中，常用自定义适配器，即继承于 BaseAdapter 的自定义适配器类。

（4）ListView 的常用 UI 属性：

① android:divider。

② android:dividerHeight。

③ android:entries。

④ android:footerDividersEnabled。

⑤ android:headerDividersEnabled。

 任务实现

具体操作步骤如下。

① 登录唯众物联网融合平台，创建项目，在项目下添加设备，添加成功后单击【生成 API】按钮。

② 在 Android Studio 中创建一个新项目,将应用名称设置为 Android-SQLiteWD-History,并为活动添加一个空活动。

③ 创建 listview_impl.xml,代码如下。

```xml
<LinearLayout
    android:layout_width="match_parent"
    android:layout_height="wrap_content"
    android:orientation="horizontal"
    android:weightSum="1">

    <TextView
        android:id="@+id/id_tv"
        android:layout_width="1dp"
        android:layout_height="wrap_content"
        android:layout_weight="0.15"
        android:textAlignment="center"/>

    <EditText
        android:id="@+id/value_et"
        android:layout_width="1dp"
        android:layout_height="wrap_content"
        android:layout_weight="0.15"
        android:textAlignment="center"/>

    <TextView
        android:id="@+id/time_tv"
        android:layout_width="1dp"
        android:layout_height="wrap_content"
        android:layout_weight="0.3"
        android:textAlignment="center"/>

    <Button
        android:id="@+id/update_bt"
        android:layout_width="1dp"
        android:layout_height="wrap_content"
        android:layout_weight="0.2"
        android:onClick="onClick"
        android:text=" 修改 "/>

    <Button
        android:id="@+id/delete_bt"
        android:layout_width="1dp"
        android:layout_height="wrap_content"
        android:layout_weight="0.2"
        android:onClick="onClick"
        android:text=" 删除 "/>
</LinearLayout>
```

④ 修改 activity_main.xml,在主活动布局中添加控件,代码如下,实现效果如图 5-3 所示。

```xml
<TextView
    android:layout_width="match_parent"
```

```xml
        android:layout_height="wrap_content"
        android:text="室内环境数据存储"
        android:textAlignment="center"
        android:textSize="25dp"/>

    <LinearLayout
        android:layout_width="match_parent"
        android:layout_height="wrap_content"
        android:orientation="horizontal">

        <LinearLayout
            android:layout_width="wrap_content"
            android:layout_height="wrap_content">

            <TextView
                android:layout_width="wrap_content"
                android:layout_height="wrap_content"
                android:text="每2秒获取的信息 ==>"
                android:textAlignment="textStart"/>
        </LinearLayout>

        <LinearLayout
            android:layout_width="match_parent"
            android:layout_height="wrap_content"
            android:orientation="horizontal">

            <TextView
                android:id="@+id/value_tv"
                android:layout_width="1dp"
                android:layout_height="wrap_content"
                android:layout_weight="0.4"
                android:text="温度值:"
                android:textAlignment="textStart"/>

            <TextView
                android:id="@+id/createTime_tv"
                android:layout_width="1dp"
                android:layout_height="wrap_content"
                android:layout_weight="0.6"
                android:text="采集时间:"
                android:textAlignment="textStart"/>
        </LinearLayout>
    </LinearLayout>

    <Button
        android:id="@+id/showData_bt"
        android:layout_width="match_parent"
        android:layout_height="wrap_content"
        android:onClick="onClick"
        android:text="点击加载采集到的数据"/>

    <Button
        android:id="@+id/toHistory_bt"
        android:layout_width="match_parent"
```

```xml
        android:layout_height="wrap_content"
        android:onClick="onClick"
        android:text=" 室内环境数据历史查询 "/>

    <LinearLayout
        android:id="@+id/title_layoutLL"
        android:layout_width="match_parent"
        android:layout_height="wrap_content"
        android:orientation="horizontal"
        android:weightSum="1">

        <TextView
            android:layout_width="1dp"
            android:layout_height="wrap_content"
            android:layout_weight="0.15"
            android:text=" 数据 id"
            android:textAlignment="center"/>

        <TextView
            android:layout_width="1dp"
            android:layout_height="wrap_content"
            android:layout_weight="0.15"
            android:text=" 温度值 "
            android:textAlignment="center"/>

        <TextView
            android:layout_width="1dp"
            android:layout_height="wrap_content"
            android:layout_weight="0.3"
            android:text=" 采集时间 "
            android:textAlignment="center"/>

        <TextView
            android:layout_width="1dp"
            android:layout_height="wrap_content"
            android:layout_weight="0.4"
            android:text=" 操作 "
            android:textAlignment="center"/>

    </LinearLayout>

    <ListView
        android:id="@+id/temp_lv"
        android:layout_width="match_parent"
        android:layout_height="wrap_content">

    </ListView>
```

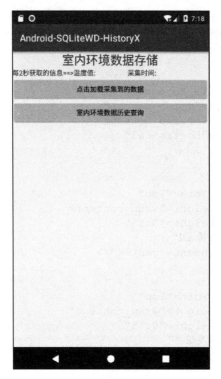

图 5-3　效果图

⑤ 创建 DateUtils.java 类，代码如下。

```
public class DateUtils {
    public static String getTodayDateTime() {
        SimpleDateFormat format=new SimpleDateFormat("yyyy-MM-dd HH:mm:ss",
            Locale.getDefault());
        return format.format(new Date());
    }

    /**
     * 调用此方法输入所要转换的时间，例如输入（"2014年06月14日16时09分00秒"）返回时间戳
     *
     * @param time
     * @return
     */
    public static String translateDate(String time) {
        SimpleDateFormat sdr=new SimpleDateFormat("yyyy年MM月dd日HH时mm分ss秒",
            Locale.CHINA);
        Date date;
        String times=null;
        try {
            date=sdr.parse(time);
            long l=date.getTime();
            times=String.valueOf(l);
        } catch (Exception e) {
            e.printStackTrace();
```

```java
        }
        return times;
    }

    /**
     * 调用此方法输入所要转换的时间戳，例如输入（"2014-06-14 16:09:00"）返回时间戳
     *
     * @param time
     * @return
     */
    public static String dateOne(String time) {
        SimpleDateFormat sdr=new SimpleDateFormat("yyyy-MM-dd HH:mm:ss",
                Locale.CHINA);
        Date date;
        String times=null;
        try {
            date=sdr.parse(time);
            long l=date.getTime();
            String stf=String.valueOf(l);
            times=stf.substring(0, 10);
        } catch (Exception e) {
            e.printStackTrace();
        }
        return times;
    }

    public static String getTimestamp(String time, String type) {
        SimpleDateFormat sdr=new SimpleDateFormat(type, Locale.CHINA);
        Date date;
        String times=null;
        try {
            date=sdr.parse(time);
            long l=date.getTime();
            String stf=String.valueOf(l);
            times=stf.substring(0, 10);
        } catch (Exception e) {
            e.printStackTrace();
        }
        return times;
    }

    /**
     * 调用此方法输入所要转换的时间戳，例如输入（1402733340）输出（"2014年06月14日16时09分00秒"）
     *
     * @param time
     * @return
     */
    public static String times(String time) {
        SimpleDateFormat sdr=new SimpleDateFormat("yyyy年MM月dd日HH时mm分ss秒");
        @SuppressWarnings("unused")
        long lcc=Long.valueOf(time);
        int i=Integer.parseInt(time);
        String times=sdr.format(new Date(i*1000L));
```

```
        return times;
    }

    /**
     * 调用此方法输入所要转换的时间戳,例如输入(1402733340)输出("2014-06-14  16:09:00")
     *
     * @param time
     * @return
     */
    public static String timedate(String time) {
        SimpleDateFormat sdr=new SimpleDateFormat("yyyy-MM-dd HH:mm:ss");
        @SuppressWarnings("unused")
        long lcc=Long.valueOf(time);
        int i=Integer.parseInt(time);
        String times=sdr.format(new Date(i * 1000L));
        return times;
    }

    /**
     * 调用此方法输入所要转换的时间戳,例如输入(1402733340)输出("2014年06月14日16:09")
     *
     * @param time
     * @return
     */
    public static String timet(String time) {
        SimpleDateFormat sdr=new SimpleDateFormat("yyyy年MM月dd日  HH:mm");
        @SuppressWarnings("unused")
        long lcc=Long.valueOf(time);
        int i=Integer.parseInt(time);
        String times=sdr.format(new Date(i*1000L));
        return times;
    }
}
```

⑥ 创建 MyAdapter.java 类,代码如下。

```
public class MyAdapter extends ArrayAdapter<Temperature> {
    public MyAdapter(Context context, int resource, List objects) {
        super(context, resource, objects);
    }
    @Override
    public View getView(int position, View convertView, ViewGroup parent) {
        Temperature temperature=(Temperature) getItem(position);
        View view=LayoutInflater.from(getContext()).inflate(R.layout.listview_impl, null);
        // 获取组件
        TextView id_tv=(TextView) view.findViewById(R.id.id_tv);
        TextView time_tv=(TextView) view.findViewById(R.id.time_tv);
        final EditText value_et=(EditText) view.findViewById(R.id.value_et);
        Button update=view.findViewById(R.id.update_bt);
```

```java
        Button delete=view.findViewById(R.id.delete_bt);

        //设置值
        id_tv.setText(temperature.getId().toString());
        value_et.setText(temperature.getValue());
        time_tv.setText(temperature.getCreateTime());

        final Integer tempId=temperature.getId();

        //设置点击事件
        update.setOnClickListener(new View.OnClickListener() {
            @Override
            public void onClick(View v) {
                SQLiteDatabase db=(SQLiteDatabase) MainActivity.dbMap.get("db");
                Handler handler=(Handler) MainActivity.dbMap.get("handler");
                ContentValues cv=new ContentValues();
                cv.put("value", value_et.getText().toString());
                int resultRow=db.update("temperature", cv, "id=?", new String[]{tempId.toString()});
                if (resultRow>0) {
                    Message message=new Message();
                    message.what=2;
                    message.obj="修改id=" + tempId + "的数据成功!";
                    handler.sendMessage(message);
                }
            }
        });
        //设置点击事件
        delete.setOnClickListener(new View.OnClickListener() {
            @Override
            public void onClick(View v) {
                SQLiteDatabase db=(SQLiteDatabase) MainActivity.dbMap.get("db");
                Handler handler=(Handler) MainActivity.dbMap.get("handler");
                int resultRow=db.delete("temperature", "id=?", new String[]{tempId.toString()});
                if (resultRow>0) {
                    Message message=new Message();
                    message.what=2;
                    message.obj="删除id=" + tempId + "的数据成功!";
                    handler.sendMessage(message);
                }
            }
        });
        return view;
    }
}
```

⑦ 创建 MyDataBaseUtil.java 类，代码如下。

```java
public class MyDataBaseUtil extends SQLiteOpenHelper {
    public MyDataBaseUtil(Context context, String databaseName, SQLiteDatabase.CursorFactory factory, int version) {
        super(context, databaseName, factory, version);
```

```java
    }

    @Override
    public void onCreate(SQLiteDatabase db) {
        // 此方法在第一次创建数据库时执行 ( 是否执行根据    databaseName.db   文件存在与否决定 )
        /*String create="CREATE TABLE temperature(id INTEGER PRIMARY KEY AUTOINCREMENT,value VARCHAR(200),createTime VARCHAR(200));";
        db.execSQL(create);*/
    }

    @Override
    public void onUpgrade(SQLiteDatabase db, int oldVersion, int newVersion) {

    }
}
```

⑧ 修改 MainActivity.java，实现效果，代码如下。

```java
public static HashMap<String, Object> dbMap=new HashMap<>();

private SQLiteDatabase db;

private TextView value_tv;
private TextView createTime_tv;
private Button showData_bt;
private LinearLayout title_layoutLL;

private SimpleDateFormat sdf=new SimpleDateFormat("yyyy/MM/dd HH:mm:ss");

@Override
protected void onCreate(Bundle savedInstanceState) {
    super.onCreate(savedInstanceState);
    setContentView(R.layout.activity_main);

    //1. 初始化组件
    initView();

    //2. 初始化数据库
    initDataBase();

    //3. 初始化请求
    initTools();
    //4.ListView列表渲染
    initListView();
}
private void initView() {
    value_tv=(TextView) findViewById(R.id.value_tv);
    createTime_tv=(TextView) findViewById(R.id.createTime_tv);
    title_layoutLL=(LinearLayout) findViewById(R.id.title_layoutLL);
    showData_bt=(Button) findViewById(R.id.showData_bt);
}

private final Timer timer=new Timer();
```

```java
Handler handler=new Handler() {
    @Override
    public void handleMessage(Message msg) {
        String jsonResult=msg.obj.toString();
        switch (msg.what) {
            case 1:
                saveData(jsonResult,"1");//index:通道号,通常为1
                break;
            case 2:
                Toast.makeText(MainActivity.this, msg.obj.toString(), Toast.LENGTH_SHORT).show();
                initListView();
                break;
            case 3:
                Toast.makeText(MainActivity.this, msg.obj.toString(), Toast.LENGTH_SHORT).show();
                break;
        }
        super.handleMessage(msg);
    }
};

private void initDataBase() {
    MyDataBaseUtil dataBaseUtil=new MyDataBaseUtil(MainActivity.this, "wd_db", null, 1);

    //每次启动项目就重新创建表,根据自身需求决定
    String drop="DROP TABLE IF EXISTS 'temperature'";
    db=dataBaseUtil.getReadableDatabase();
    db.execSQL(drop);

    //创建表,如果不需要每次启动项目重新创建,可将该段代码添加到MyDataBaseUtil.onCreate中
    String create="CREATE TABLE temperature(id INTEGER PRIMARY KEY AUTOINCREMENT,value VARCHAR(200),createTime VARCHAR(200));";
    db.execSQL(create);

    //getReadableDatabase()及getWritableDatabase在数据库第一次创建时将会调用MyDataBaseUtil.onCreate方法
    db=dataBaseUtil.getWritableDatabase();
    dbMap.put("db", db);
    dbMap.put("handler", handler);
}

private void initTools() {
    Wz_HttpTools wht=new Wz_HttpTools(handler);
    wht.setHttpURL("http://192.168.0.193:8080/wziot/wzIotApi/getOneSensorData/3179a728-51c4-4fcc-9454-d7324c72187d/30008");
    TimerTask task=wht.getJsonData();
    timer.schedule(task, 2000, 5000);//每隔2s采集一次数据
}

//将采集到的数据进行存储
private void saveData(String jsonResult, String index) {
    try {
```

```java
                // 解析json串，获取信息内容部分
                JSONArray obj=new JSONObject(jsonResult).getJSONArray("res");
                // 循环提取需要的通道号
                for (int i=0; i<obj.length(); i++) {
                    JSONObject json=(JSONObject) obj.get(i);
                    String rindex=json.getString("passGatewayNum");
                    if (index.equals(rindex)) {// 获取温度值
                        String time=json.getString("time");
                        String value=json.getString("value");
                        ContentValues cv=new ContentValues();
                        cv.put("value", value);
                        cv.put("createTime", time);
                        value_tv.setText("温度值:" + value);
                        createTime_tv.setText("采集时间:" + time);
                        long resultRow=db.insert("temperature", null, cv);
                        /*if (resultRow>0) {
                            Message message=new Message();
                            message.what=3;
                            message.obj=" 采集到了一条数据 \n温度值为:" + value + "\n采
集时间为:"+sdf.format(Long.valueOf(time));
                            handler.sendMessage(message);
                        }*/
                    }
                }
            } catch (JSONException e) {
                e.printStackTrace();
            }
        }
    }

    private void initListView() {
        List<Temperature> temperatureList=new ArrayList<Temperature>();
        Cursor cursor;
        // 获取数据添加到集合中
        cursor=db.query("temperature", new String[]{"id", "value",
"createTime"}, null, null, null, null, "id desc");
        while (cursor.moveToNext()) {
            Integer id=cursor.getInt(cursor.getColumnIndex("id"));
            String value=cursor.getString(cursor.getColumnIndex("value"));
            String createTime=cursor.getString(cursor.
getColumnIndex("createTime"));
            temperatureList.add(new Temperature(id, value, sdf.format(Long.
valueOf(createTime))));
        }
        if (temperatureList.size()<=0) {
            title_layoutLL.setVisibility(View.GONE);
        } else {
            title_layoutLL.setVisibility(View.VISIBLE);
        }
        // 初始化适配器并进行ListView UI 渲染
        MyAdapter adapter=new MyAdapter(MainActivity.this, R.layout.listview_
impl, temperatureList);
        ListView listView=(ListView) findViewById(R.id.temp_lv);
        listView.setAdapter(adapter);
    }
```

```
@Override
public void onClick(View v) {
    switch (v.getId()) {
        case R.id.showData_bt: {
            initListView();
            break;
        }
        case R.id.toHistory_bt: {
            startActivity(new Intent(MainActivity.this, HistoryActivity.class));
            finish();
            break;
        }
    }
}
```

 任务小结

请同学们根据完成情况对完成本次任务的知识、技能等要点进行小结。

任务 2 小结	
知识点掌握情况	
技能点掌握情况	

任务拓展　使用 ListView 显示数据

 任务描述

数据较多时可以用列表的形式显示数据。

 **任务分析**

使用 ListView 来显示数据。

 任务实现

具体操作步骤如下。

① 登录唯众物联网融合平台，创建项目，在项目下添加设备，添加成功后单击【生成 API】按钮。

② 本项目在上一个项目基础上完成。

③ 创建 lactivity_history.xml，代码如下。

```xml
<TextView
    android:layout_width="match_parent"
    android:layout_height="wrap_content"
    android:text="室内环境数据历史查询"
    android:textAlignment="center"
    android:textSize="25dp" />

<LinearLayout
    android:layout_width="match_parent"
    android:layout_height="wrap_content"
    android:orientation="horizontal">

    <Button
        android:id="@+id/startDate_btn"
        android:layout_width="wrap_content"
        android:layout_height="wrap_content"
        android:onClick="showDatePickerDialog"
        android:text="请选择开始日期" />

    <Button
        android:id="@+id/startTime_btn"
        android:layout_width="wrap_content"
        android:layout_height="wrap_content"
        android:onClick="showTimePickerDialog"
        android:text="请选择开始时间" />
</LinearLayout>

<LinearLayout
    android:layout_width="match_parent"
    android:layout_height="wrap_content"
    android:orientation="horizontal">

    <TextView
        android:layout_width="wrap_content"
        android:layout_height="wrap_content"
        android:text="你选择的开始时间为:" />

    <TextView
        android:id="@+id/startDate_tv"
        android:layout_width="wrap_content"
        android:layout_height="wrap_content"
        android:text="" />

    <TextView
        android:id="@+id/startTime_tv"
        android:layout_width="wrap_content"
        android:layout_height="wrap_content"
        android:text="" />
</LinearLayout>

<LinearLayout
    android:layout_width="match_parent"
    android:layout_height="wrap_content"
    android:orientation="horizontal">
```

```xml
    <Button
        android:id="@+id/endDate_btn"
        android:layout_width="wrap_content"
        android:layout_height="wrap_content"
        android:onClick="showDatePickerDialog"
        android:text="请选择结束日期" />

    <Button
        android:id="@+id/endTime_btn"
        android:layout_width="wrap_content"
        android:layout_height="wrap_content"
        android:onClick="showTimePickerDialog"
        android:text="请选择结束时间" />
</LinearLayout>

<LinearLayout
    android:layout_width="match_parent"
    android:layout_height="wrap_content"
    android:orientation="horizontal">

    <TextView
        android:layout_width="wrap_content"
        android:layout_height="wrap_content"
        android:text="你选择的结束时间为:" />

    <TextView
        android:id="@+id/endDate_tv"
        android:layout_width="wrap_content"
        android:layout_height="wrap_content"
        android:text="" />

    <TextView
        android:id="@+id/endTime_tv"
        android:layout_width="wrap_content"
        android:layout_height="wrap_content"
        android:text="" />
</LinearLayout>

<Button
    android:layout_width="match_parent"
    android:layout_height="wrap_content"
    android:onClick="getHistoryDate"
    android:text="查询历史数据" />

<LinearLayout
    android:id="@+id/title_layoutLL"
    android:layout_width="match_parent"
    android:layout_height="wrap_content"
    android:orientation="horizontal"
    android:weightSum="1">

    <TextView
        android:layout_width="1dp"
```

```xml
        android:layout_height="wrap_content"
        android:layout_weight="0.25"
        android:text=" 数据 id"
        android:textAlignment="center" />

    <TextView
        android:layout_width="1dp"
        android:layout_height="wrap_content"
        android:layout_weight="0.25"
        android:text=" 温度值 "
        android:textAlignment="center" />

    <TextView
        android:layout_width="1dp"
        android:layout_height="wrap_content"
        android:layout_weight="0.5"
        android:text=" 采集时间 "
        android:textAlignment="center" />

</LinearLayout>

<ListView
    android:id="@+id/history_lv"
    android:layout_width="match_parent"
    android:layout_height="wrap_content"></ListView>
```

④ 创建 listview_history.xml，代码如下。

```xml
<LinearLayout
    android:layout_width="match_parent"
    android:layout_height="wrap_content"
    android:orientation="horizontal"
    android:weightSum="1">

    <TextView
        android:id="@+id/id_tv"
        android:layout_width="1dp"
        android:layout_height="wrap_content"
        android:layout_weight="0.25"
        android:textAlignment="center" />

    <EditText
        android:id="@+id/value_et"
        android:layout_width="1dp"
        android:layout_height="wrap_content"
        android:layout_weight="0.25"
        android:textAlignment="center" />

    <TextView
        android:id="@+id/time_tv"
        android:layout_width="1dp"
        android:layout_height="wrap_content"
        android:layout_weight="0.5"
```

```
        android:textAlignment="center" />
</LinearLayout>
```

⑤ 修改 activity_main.xml，代码如下。

```xml
<TextView
    android:layout_width="match_parent"
    android:layout_height="wrap_content"
    android:text="室内环境数据存储"
    android:textAlignment="center"
    android:textSize="25dp" />

<LinearLayout
    android:layout_width="match_parent"
    android:layout_height="wrap_content"
    android:orientation="horizontal">

    <LinearLayout
        android:layout_width="wrap_content"
        android:layout_height="wrap_content">

        <TextView
            android:layout_width="wrap_content"
            android:layout_height="wrap_content"
            android:text="每2s获取的信息 ==>"
            android:textAlignment="textStart" />
    </LinearLayout>

    <LinearLayout
        android:layout_width="match_parent"
        android:layout_height="wrap_content"
        android:orientation="horizontal">

        <TextView
            android:id="@+id/value_tv"
            android:layout_width="1dp"
            android:layout_height="wrap_content"
            android:layout_weight="0.4"
            android:text="温度值："
            android:textAlignment="textStart" />

        <TextView
            android:id="@+id/createTime_tv"
            android:layout_width="1dp"
            android:layout_height="wrap_content"
            android:layout_weight="0.6"
            android:text="采集时间："
            android:textAlignment="textStart" />
    </LinearLayout>
</LinearLayout>

<Button
    android:id="@+id/showData_bt"
    android:layout_width="match_parent"
```

```xml
        android:layout_height="wrap_content"
        android:onClick="onClick"
        android:text=" 点击加载采集到的数据 " />

    <Button
        android:id="@+id/toHistory_bt"
        android:layout_width="match_parent"
        android:layout_height="wrap_content"
        android:onClick="onClick"
        android:text=" 室内环境数据历史查询 " />

    <LinearLayout
        android:id="@+id/title_layoutLL"
        android:layout_width="match_parent"
        android:layout_height="wrap_content"
        android:orientation="horizontal"
        android:weightSum="1">

        <TextView
            android:layout_width="1dp"
            android:layout_height="wrap_content"
            android:layout_weight="0.15"
            android:text=" 数据 id"
            android:textAlignment="center" />

        <TextView
            android:layout_width="1dp"
            android:layout_height="wrap_content"
            android:layout_weight="0.15"
            android:text=" 温度值 "
            android:textAlignment="center" />

        <TextView
            android:layout_width="1dp"
            android:layout_height="wrap_content"
            android:layout_weight="0.3"
            android:text=" 采集时间 "
            android:textAlignment="center" />

        <TextView
            android:layout_width="1dp"
            android:layout_height="wrap_content"
            android:layout_weight="0.4"
            android:text=" 操作 "
            android:textAlignment="center" />

    </LinearLayout>

    <ListView
        android:id="@+id/temp_lv"
        android:layout_width="match_parent"
        android:layout_height="wrap_content">

    </ListView>
```

⑥ 创建 HistoryActivity.java 类，代码如下。

```java
public class HistoryActivity extends AppCompatActivity {
    private Button startDate_btn;
    private Button startTime_btn;
    private Button endDate_btn;
    private Button endTime_btn;
    private TextView startDate_tv;
    private TextView startTime_tv;
    private TextView endDate_tv;
    private TextView endTime_tv;
    private LinearLayout title_layoutLL;

    private SQLiteDatabase db;
    private SimpleDateFormat sdf=new SimpleDateFormat("yyyy/MM/dd HH:mm:ss");

    @Override
    protected void onCreate(Bundle savedInstanceState) {
        super.onCreate(savedInstanceState);
        setContentView(R.layout.activity_history);
        db=(SQLiteDatabase) MainActivity.dbMap.get("db");
        initView();
    }

    //---------------------------点击事件---------------------------------
    private void initView() {
//        startDate_btn=(Button) findViewById(R.id.startDate_btn);
//        startTime_btn=(Button) findViewById(R.id.startTime_btn);
//        endDate_btn=(Button) findViewById(R.id.endDate_btn);
//        endTime_btn=(Button) findViewById(R.id.endTime_btn);
        startDate_tv=(TextView) findViewById(R.id.startDate_tv);
        startTime_tv=(TextView) findViewById(R.id.startTime_tv);
        endDate_tv=(TextView) findViewById(R.id.endDate_tv);
        endTime_tv=(TextView) findViewById(R.id.endTime_tv);
        title_layoutLL=(LinearLayout) findViewById(R.id.title_layoutLL);

    }

    public void showDatePickerDialog(View view) {
        String msg="";
        boolean flag=false;
        if (R.id.startDate_btn==view.getId()) {
            msg="开始日期为:";
            flag=true;
        } else if (R.id.endDate_btn==view.getId()) {
            msg="结束日期为:";
            flag=false;
        }
        final String dateMsg=msg;
        final boolean dateFlag=flag;
        DatePickerDialog datePickerDialog=new DatePickerDialog(HistoryActivity.this, new DatePickerDialog.OnDateSetListener() {
            @Override
```

```java
            public void onDateSet(DatePicker view, int year, int month, int day) {
                String yearStr=year + "年";
                String monthStr=(month + 1)<10? "0"+(month + 1)+"月" : (month + 1) + "月";
                String dayStr=day<10? "0"+day+"日" : day+"日";
                String dateInfo=yearStr + monthStr + dayStr;
                if (dateFlag) {
                    startDate_tv.setText(dateInfo);
                } else {
                    endDate_tv.setText(dateInfo);
                }
                Toast.makeText(HistoryActivity.this, dateMsg + dateInfo, Toast.LENGTH_SHORT).show();
            }
        }, 2019, 4, 28);
        datePickerDialog.show();
    }

    public void showTimePickerDialog(View view) {
        String msg="";
        boolean flag=false;
        if (R.id.startTime_btn==view.getId()) {
            msg="开始时分为:";
            flag=true;
        } else if (R.id.endTime_btn==view.getId()) {
            msg="结束时分为:";
            flag=false;
        }
        final String timeMsg=msg;
        final boolean timeFlag=flag;
        TimePickerDialog time=new TimePickerDialog(HistoryActivity.this, new TimePickerDialog.OnTimeSetListener() {
            @Override
            public void onTimeSet(TimePicker view, int hour, int minute) {
                String hourStr=hour<10? "0"+hour+"时" : hour+"时";
                String minuteStr=minute<10? "0"+minute+"分" : minute+"分";
                String second="00秒";
                String dateInfo=hourStr+minuteStr+second;
                if (timeFlag) {
                    startTime_tv.setText(dateInfo);
                } else {
                    endTime_tv.setText(dateInfo);
                }
                Toast.makeText(HistoryActivity.this, timeMsg+dateInfo, Toast.LENGTH_SHORT).show();
            }
        }, 15, 43, true);
        time.show();
    }

    public void getHistoryDate(View view) {
        String startDate=startDate_tv.getText().toString();
        String endDate=endDate_tv.getText().toString();
        String startTime=startTime_tv.getText().toString();
```

```java
            String endTime=endTime_tv.getText().toString();
            if ("".equals(startDate.trim())) {
                Toast.makeText(HistoryActivity.this, "请选择开始日期!", Toast.
LENGTH_SHORT).show();
                return;
            } else if ("".equals(endDate.trim())) {
                Toast.makeText(HistoryActivity.this, "请选择结束日期!", Toast.
LENGTH_SHORT).show();
                return;
            } else if ("".equals(startTime.trim())) {
                Toast.makeText(HistoryActivity.this, "请选择开始时分!", Toast.
LENGTH_SHORT).show();
                return;
            } else if ("".equals(endTime.trim())) {
                Toast.makeText(HistoryActivity.this, "请选择结束时分!", Toast.
LENGTH_SHORT).show();
                return;
            }
            String start=startDate + startTime;
            String end=endDate + endTime;
            String time1=DateUtils.translateDate(start);
            String time2=DateUtils.translateDate(end);
            if (time1.equals(time2)) {
                Toast.makeText(HistoryActivity.this, "开始时间与结束时间不能相同!",
Toast.LENGTH_SHORT).show();
                return;
            }
            int result=Long.compare(Long.parseLong(time1), Long.
parseLong(time2));
            String trans="";
            if (result>0) {
                trans=startDate;
                startDate=endDate;
                endDate=trans;

                trans=startTime;
                startTime=endTime;
                endTime=trans;

                trans=time1;
                time1=time2;
                time2=trans;

                startDate_tv.setText(startDate);
                endDate_tv.setText(endDate);
                startTime_tv.setText(startTime);
                endTime_tv.setText(endTime);
                Toast.makeText(HistoryActivity.this, "检测到开始时间比结束时间大,已自
动进行调换!", Toast.LENGTH_SHORT).show();
            }
            initListView(time1, time2);
        }
```

```java
//-------------------------------- 点击事件 ---------------------------------
    private Handler handler=new Handler() {
        @Override
        public void handleMessage(Message msg) {
            switch (msg.what) {
                case 1: {
                    break;
                }
            }
            super.handleMessage(msg);
        }
    };


    // 加载ListView列表
    private void initListView(String startTime, String endTime) {
        List<Temperature> temperatureList=new ArrayList<Temperature>();
        Cursor cursor;
        // 获取数据添加到集合中
        cursor=db.query("temperature", new String[]{"id", "value", "createTime"}, "createTime>=" + startTime + " AND createTime<=" + endTime, null, null, null, "id desc");
        while (cursor.moveToNext()) {
            Integer id=cursor.getInt(cursor.getColumnIndex("id"));
            String value=cursor.getString(cursor.getColumnIndex("value"));
            String createTime=cursor.getString(cursor.getColumnIndex("createTime"));
            temperatureList.add(new Temperature(id, value, sdf.format(Long.valueOf(createTime))));
        }
        if (temperatureList.size()<=0) {
            title_layoutLL.setVisibility(View.GONE);
        } else {
            title_layoutLL.setVisibility(View.VISIBLE);
        }
        // 初始化适配器并进行ListView UI渲染
        HistoryAdapter adapter=new HistoryAdapter(HistoryActivity.this, R.layout.listview_history, temperatureList);
        ListView listView=(ListView) findViewById(R.id.history_lv);
        listView.setAdapter(adapter);
    }


}
```

⑦ 创建HistoryAdapter.java类，代码如下。

```java
public class HistoryAdapter extends ArrayAdapter<Temperature> {
    public HistoryAdapter(Context context, int resource, List objects) {
        super(context, resource, objects);
    }
```

```java
        @Override
        public View getView(int position, View convertView, ViewGroup parent) {
            Temperature temperature=(Temperature) getItem(position);
            View view=LayoutInflater.from(getContext()).inflate(R.layout.listview_history, null);

            // 获取组件
            TextView id_tv=(TextView) view.findViewById(R.id.id_tv);
            TextView time_tv=(TextView) view.findViewById(R.id.time_tv);
            EditText value_et=(EditText) view.findViewById(R.id.value_et);

            // 设置值
            id_tv.setText(temperature.getId().toString());
            value_et.setText(temperature.getValue());
            time_tv.setText(temperature.getCreateTime());
            return view;
        }
    }
```

⑧ 修改 MainActivity.java，实现效果，代码如下。

```java
public class MainActivity extends AppCompatActivity implements View.OnClickListener {
    public static HashMap<String, Object> dbMap=new HashMap<>();

    private SQLiteDatabase db;

    private TextView value_tv;
    private TextView createTime_tv;
    private Button showData_bt;
    private LinearLayout title_layoutLL;

    private SimpleDateFormat sdf=new SimpleDateFormat("yyyy/MM/dd HH:mm:ss");

    @Override
    protected void onCreate(Bundle savedInstanceState) {
        super.onCreate(savedInstanceState);
        setContentView(R.layout.activity_main);

        //1. 初始化组件
        initView();

        //2. 初始化数据库
        initDataBase();

        //3. 初始化请求
        initTools();

        //4.ListView 列表渲染
        initListView();
    }

    private void initView() {
        value_tv=(TextView) findViewById(R.id.value_tv);
```

```java
            createTime_tv=(TextView) findViewById(R.id.createTime_tv);
            title_layoutLL=(LinearLayout) findViewById(R.id.title_layoutLL);
            showData_bt=(Button) findViewById(R.id.showData_bt);
    }

    private final Timer timer=new Timer();
    Handler handler=new Handler() {
        @Override
        public void handleMessage(Message msg) {
            String jsonResult=msg.obj.toString();
            switch (msg.what) {
                case 1:
                    saveData(jsonResult, "1");//index:通道号,通常为1
                    break;
                case 2:
                    Toast.makeText(MainActivity.this, msg.obj.toString(),
Toast.LENGTH_SHORT).show();
                    initListView();
                    break;
                case 3:
                    Toast.makeText(MainActivity.this, msg.obj.toString(),
Toast.LENGTH_SHORT).show();
                    break;
            }
            super.handleMessage(msg);
        }
    };

    private void initDataBase() {
        MyDataBaseUtil dataBaseUtil=new MyDataBaseUtil(MainActivity.this,
"wd_db", null, 1);

        // 每次启动项目就重新创建表,根据自身需求决定
        String drop="DROP TABLE IF EXISTS 'temperature'";
        db=dataBaseUtil.getReadableDatabase();
        db.execSQL(drop);

        // 创建表,如果不需要每次启动项目重新创建,可将该段代码添加到MyDataBaseUtil.
        //onCreate中
        String create="CREATE TABLE temperature(id INTEGER PRIMARY KEY
AUTOINCREMENT,value VARCHAR(200),createTime VARCHAR(200));";
        db.execSQL(create);

        //getReadableDatabase() 及 getWritableDatabase在数据库第一次创建时将会调用
        //MyDataBaseUtil.onCreate方法
        db=dataBaseUtil.getWritableDatabase();
        dbMap.put("db", db);
        dbMap.put("handler", handler);
    }

    private void initTools() {
        Wz_HttpTools wht=new Wz_HttpTools(handler);
        wht.setHttpURL("http://192.168.0.193:8080/wziot/wzIotApi/
getOneSensorData/3179a728-51c4-4fcc-9454-d7324c72187d/30008");
```

```
            TimerTask task=wht.getJsonData();
            timer.schedule(task, 2000, 5000);//每隔两秒采集一次数据
    }

    //将采集到的数据进行存储
    private void saveData(String jsonResult, String index) {
        try {
            //解析json串，获取信息内容部分
            JSONArray obj=new JSONObject(jsonResult).getJSONArray("res");
            //循环提取需要的通道号
            for (int i=0; i<obj.length(); i++) {
                JSONObject json=(JSONObject) obj.get(i);
                String rindex=json.getString("passGatewayNum");
                if (index.equals(rindex)) {//获取温度值
                    String time=json.getString("time");
                    String value=json.getString("value");
                    ContentValues cv=new ContentValues();
                    cv.put("value", value);
                    cv.put("createTime", time);
                    value_tv.setText("温度值:" + value);
                    createTime_tv.setText("采集时间:" + time);
                    long resultRow=db.insert("temperature", null, cv);
                    /*if (resultRow>0) {
                        Message message=new Message();
                        message.what=3;
                        message.obj="采集到了一条数据\n温度值为:" + value +
"\n采集时间为:" + sdf.format(Long.valueOf(time));
                        handler.sendMessage(message);
                    }*/
                }
            }
        } catch (JSONException e) {
            e.printStackTrace();
        }
    }

    private void initListView() {
        List<Temperature> temperatureList=new ArrayList<Temperature>();
        Cursor cursor;
        //获取数据添加到集合中
        cursor=db.query("temperature", new String[]{"id", "value",
"createTime"}, null, null, null, null, "id desc");
        while (cursor.moveToNext()) {
            Integer id=cursor.getInt(cursor.getColumnIndex("id"));
            String value=cursor.getString(cursor.getColumnIndex("value"));
            String createTime=cursor.getString(cursor.
getColumnIndex("createTime"));
            temperatureList.add(new Temperature(id, value, sdf.format(Long.
valueOf(createTime))));
        }
        if (temperatureList.size()<=0) {
            title_layoutLL.setVisibility(View.GONE);
        } else {
            title_layoutLL.setVisibility(View.VISIBLE);
```

```java
        }
        // 初始化适配器并进行 ListView UI 渲染
        MyAdapter adapter=new MyAdapter(MainActivity.this, R.layout.listview_impl, temperatureList);
        ListView listView=(ListView) findViewById(R.id.temp_lv);
        listView.setAdapter(adapter);
    }

    @Override
    public void onClick(View v) {
        switch (v.getId()) {
            case R.id.showData_bt: {
                initListView();
                break;
            }
            case R.id.toHistory_bt: {
                startActivity(new Intent(MainActivity.this, HistoryActivity.class));
                finish();
                break;
            }
        }
    }
}
```